Standard Grade | Credit

Physics

Credit Level 2001
_____

Credit Level 2002
_____

Credit Level 2003
_____

Credit Level 2004
_____

Credit Level 2005
_____

Leckie ✕ Leckie

First exam published in 2001.
Published by Leckie & Leckie, 8 Whitehill Terrace, St. Andrews, Scotland KY16 8RN tel: 01334 475656 fax: 01334 477392
enquiries@leckieandleckie.co.uk www.leckieandleckie.co.uk

ISBN 1-84372-311-5

A CIP Catalogue record for this book is available from the British Library.

Printed in Scotland by Scotprint.

Leckie & Leckie is a division of Granada Learning Limited, part of ITV plc.

Acknowledgements

Leckie & Leckie is grateful to the copyright holders, as credited at the back of the book, for permission to use their material.
Every effort has been made to trace the copyright holders and to obtain their permission for the use of copyright material.
Leckie & Leckie will gladly receive information enabling them to rectify any error or omission in subsequent editions.

[BLANK PAGE]

FOR OFFICIAL USE

C

K & U   PS

Total Marks

3220/402

NATIONAL
QUALIFICATIONS
2001

MONDAY, 4 JUNE
10.50 AM – 12.35 PM

PHYSICS
STANDARD GRADE
Credit Level

Fill in these boxes and read what is printed below.

Full name of centre

Town

Forename(s)

Surname

Date of birth
Day Month Year

Scottish candidate number

Number of seat

1 All questions should be answered.

2 The questions may be answered in any order but all answers must be written clearly and legibly in this book.

3 Write your answer where indicated by the question or in the space provided after the question.

4 If you change your mind about your answer you may score it out and rewrite it in the space provided at the end of the answer book.

5 Before leaving the examination room you must give this book to the invigilator. If you do not, you may lose all the marks for this paper.

6 Any necessary data will be found in the **data sheet** on page two.

SCOTTISH
QUALIFICATIONS
AUTHORITY

DATA SHEET

Speed of light in materials

Material	Speed in m/s
Air	$3 \cdot 0 \times 10^8$
Carbon dioxide	$3 \cdot 0 \times 10^8$
Diamond	$1 \cdot 2 \times 10^8$
Glass	$2 \cdot 0 \times 10^8$
Glycerol	$2 \cdot 1 \times 10^8$
Water	$2 \cdot 3 \times 10^8$

Speed of sound in materials

Material	Speed in m/s
Aluminium	5200
Air	340
Bone	4100
Carbon dioxide	270
Glycerol	1900
Muscle	1600
Steel	5200
Tissue	1500
Water	1500

Gravitational field strengths

	Gravitational field strength on the surface in N/kg
Earth	10
Jupiter	26
Mars	4
Mercury	4
Moon	$1 \cdot 6$
Neptune	12
Saturn	11
Sun	270
Venus	9

Specific heat capacity of materials

Material	Specific heat capacity in J/kg °C
Alcohol	2350
Aluminium	902
Copper	386
Diamond	530
Glass	500
Glycerol	2400
Ice	2100
Lead	128
Water	4180

Specific latent heat of fusion of materials

Material	Specific latent heat of fusion in J/kg
Alcohol	$0 \cdot 99 \times 10^5$
Aluminium	$3 \cdot 95 \times 10^5$
Carbon dioxide	$1 \cdot 80 \times 10^5$
Copper	$2 \cdot 05 \times 10^5$
Glycerol	$1 \cdot 81 \times 10^5$
Lead	$0 \cdot 25 \times 10^5$
Water	$3 \cdot 34 \times 10^5$

Melting and boiling points of materials

Material	Melting point in °C	Boiling point in °C
Alcohol	−98	65
Aluminium	660	2470
Copper	1077	2567
Glycerol	18	290
Lead	328	1737
Turpentine	−10	156

Specific latent heat of vaporisation of materials

Material	Specific latent heat of vaporisation in J/kg
Alcohol	$11 \cdot 2 \times 10^5$
Carbon dioxide	$3 \cdot 77 \times 10^5$
Glycerol	$8 \cdot 30 \times 10^5$
Turpentine	$2 \cdot 90 \times 10^5$
Water	$22 \cdot 6 \times 10^5$

SI Prefixes and Multiplication Factors

Prefix	Symbol	Factor	
giga	G	1 000 000 000	$= 10^9$
mega	M	1 000 000	$= 10^6$
kilo	k	1000	$= 10^3$
milli	m	0·001	$= 10^{-3}$
micro	μ	0·000 001	$= 10^{-6}$
nano	n	0·000 000 001	$= 10^{-9}$

Marks | K&U | PS

1. The depth of the seabed is measured using pulses of ultrasound waves. The ultrasound waves are transmitted from a stationary ship. The waves are reflected from the seabed as shown and are detected by equipment on the ship. The transmitted ultrasound waves have a frequency of 30 kHz.

transmitted pulse

reflected pulse   seabed

(a) One pulse of ultrasound waves is received back at the ship 0·2 s after being sent out.

(i) Use the data sheet to find the speed of the ultrasound waves in the water.

.... **1**

(ii) Calculate the depth of the seabed.

Space for working and answer

3

(iii) Calculate the wavelength of the ultrasound waves in the water.

Space for working and answer

2

[Turn over

DO NOT
WRITE IN
THIS
MARGIN

Marks | K&U | PS

1. (continued)

(b) The ultrasound waves lose energy as they travel through the water. The transmitted wave is displayed on an oscilloscope screen as shown.

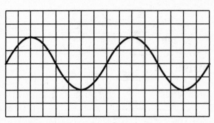

Transmitted

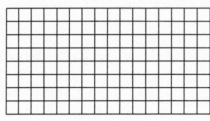

Reflected

On the bottom part of the diagram, sketch the trace produced by the reflected wave.

2

(c) The frequency of the transmitted wave is increased to 60 kHz.

What happens to the time interval between the transmitted pulse and the reflected pulse?

Explain your answer.

....

....

....

....

2

DO NOT WRITE IN THIS MARGIN

Marks | K&U | PS

2. A mobile phone has a power of 75 mW and operates using a 3 V battery.

(*a*) Calculate the current taken from the battery when the mobile phone is being used.

Space for working and answer

2

(*b*) Which of the following fuses should be connected in series with the battery of the mobile phone?

20 mA **100 mA** **2 A** **3 A**

.... 1

[Turn over

Marks | K&U | PS

3. A 2·5 V, 100 mA lamp is operated at its correct power rating from a 12 V battery by using the circuit shown.

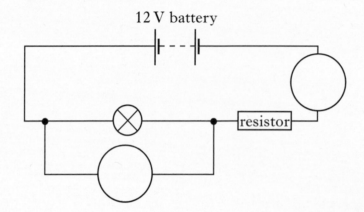

12 V battery

resistor

(a) A voltmeter and an ammeter included in the circuit show that the lamp is operating at its correct rating.

Enter the readings that are seen on the meters. Include the units for both readings.

2

(b) (i) Calculate the voltage across the resistor.

Space for working and answer

1

(ii) Calculate the resistance of the resistor.

Space for working and answer

2

4. A simple d.c. motor is shown in Figure 1.

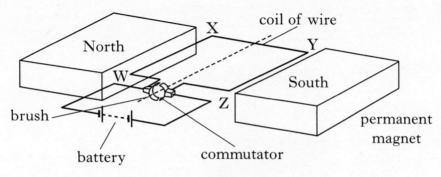

Figure 1

(a) The coil WXYZ rotates in a clockwise direction.

State **two** changes that could be made to make the coil rotate in the opposite direction.

Change 1 ....

....

Change 2 ....

.... 2

[Turn over

Official SQA Past Papers: Credit Physics 2001

DO NOT
WRITE IN
THIS
MARGIN

Marks | K&U | PS

4. (continued)

(b) Part of a commercial electric motor is shown in Figure 2.

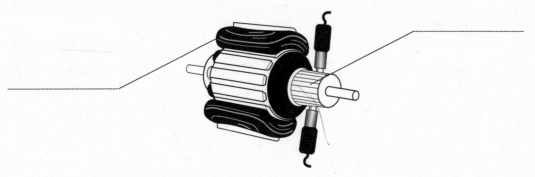

Figure 2

(i) Label the two parts indicated on the motor, using names from the list below.

brush   commutator   field coil   rotating coil   2

(ii) In the commercial electric motor, state why

(A) more than one rotating coil is used

....

....   1

(B) field coils rather than permanent magnets are used.

....

....   1

Marks | K&U | PS

5. (a) A long-sighted person is prescribed glasses that have lenses each with a power of 2·5 D.

    (i) State what is meant by long-sight.

        ....

          1

    (ii) Calculate the focal length of each lens.

        Space for working and answer

        2

(b) Complete the diagram below to show the path of the ray of light after it emerges from the lens.

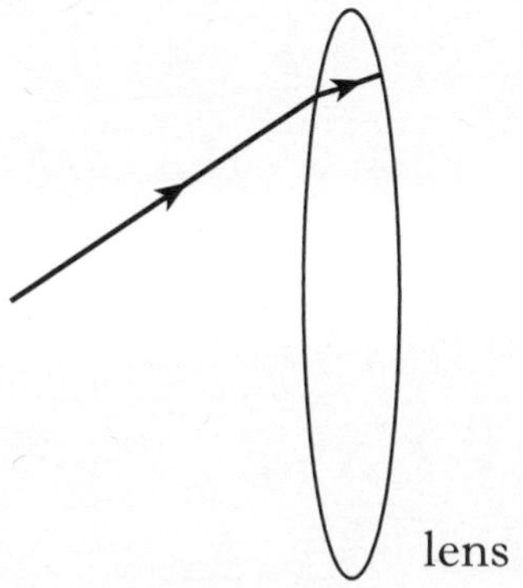

lens

1

[Turn over

Marks | K&U | PS

6. Carbon dating is used by scientists to tell the age of organic (formerly living) material. This method is based on knowing that the half-life of radioactive carbon is 5730 years.

(a) Explain what is meant by the statement "the half-life of radioactive carbon is 5730 years".

....

.... 2

(b) The proportion of radioactive carbon in the organic material is found by measuring its activity using a scintillation counter.

(i) State the **unit** that is used for the activity of a radioactive source.

.... 1

(ii) Describe how a scintillation counter is used as a detector of radiation.

....

....

.... 2

(iii) State an example of the effect of radiation other than scintillations.

.... 1

Marks | K&U | PS

7. A thermistor is used as a temperature sensor in the voltage divider circuit shown below. The circuit is used to sense the temperature of water in a tank. When the temperature of the water in the tank falls below a certain value, the output of the voltage divider causes a switching circuit to operate a heater.

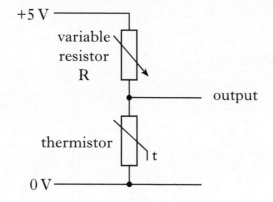

(a) When the voltage across the thermistor reaches 0·7 V, the circuit causes the heater to be switched on.

   (i) The variable resistor R is set to a resistance of 4300 Ω.

   Calculate the resistance of the thermistor when the voltage across the thermistor is 0·7 V.

   Space for working and answer

2

[Turn over

Marks | K&U | PS

7. (a) (continued)

(ii) The graph shows how the resistance of the thermistor changes with temperature.

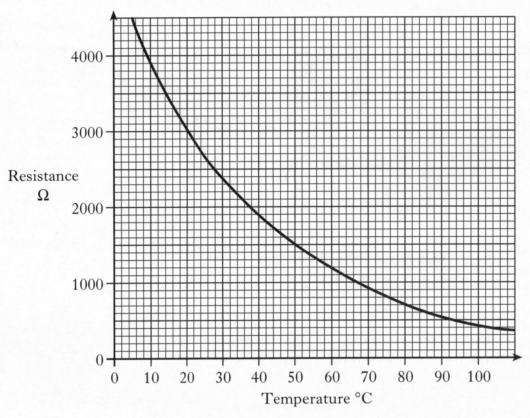

(A) Use the graph to decide the temperature at which the heater is switched on.

.... 1

(B) The resistance of the variable resistor R is increased to a value **greater than** 4300 Ω.

What effect does this have on the temperature at which the heater is switched on?

Explain your answer.

....

....

.... 2

Marks

7.   (continued)

(*b*) The voltage divider circuit is connected to the switching circuit, as shown, to operate the heater. When there is a current in the relay coil, the relay switch closes.

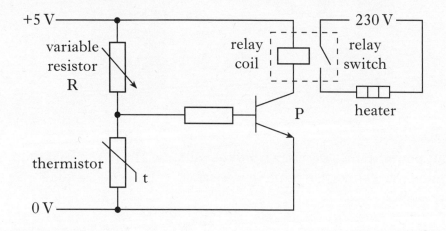

(i)   Name component P.

....   **1**

(ii) Explain why the heater switches on as the temperature falls below a selected value.

....

....

....

....

....   **3**

[Turn over

Marks | K&U | PS

8. An electric guitar is connected to an amplifier.

The input power to the amplifier from the guitar is 16 mW. The output of the amplifier is connected to a loudspeaker. The loudspeaker has a resistance of 9 Ω.

(*a*) The amplifier delivers an output power of 64 W to the loudspeaker.

(i) Calculate the power gain of the amplifier.

Space for working and answer

2

(ii) Calculate the voltage across the loudspeaker.

Space for working and answer

2

DO NOT WRITE IN THIS MARGIN

Marks | K&U | PS

8. (continued)

(*b*) A second, identical loudspeaker is connected in parallel with the first.

Calculate the combined resistance of the two loudspeakers in parallel.

Space for working and answer

2

(*c*) The guitarist plays a note of frequency 256 Hz.

What is the frequency of the output signal from the amplifier?

.... 1

[Turn over

Marks | K&U | PS

9. A cyclist has a small computer attached to her bike. The computer gives information on the cyclist's instantaneous speed, distance travelled and time taken.

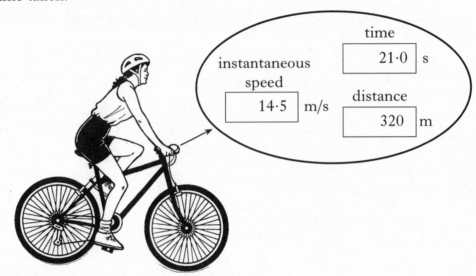

At a point during a journey, the readings on the display are as shown above.

(a) (i) Calculate the average speed of the cyclist up to this point.

(You must use an appropriate number of significant figures in your answer to this question.)

Space for working and answer

3

(ii) Why is the average speed of the cyclist not always the same as the instantaneous speed displayed on the computer?

....

....

....   2

DO NOT WRITE IN THIS MARGIN

Marks	K&U	PS

9. (continued)

(*b*) (i) The total mass of the cyclist and bike is 80 kg.

Calculate the total kinetic energy of the cyclist and the bike at this point during the journey.

Space for working and answer

2

(ii) The cyclist brakes to a halt in a distance of 50 m.

Calculate the braking force used.

Space for working and answer

2

[Turn over

Marks | K&U | PS

10. An aircraft has a mass of 268 000 kg. The aircraft accelerates from rest along a straight runway. It takes 40 s for the aircraft to reach its take-off speed of 80 m/s.

(*a*) The speed-time graph of the aircraft is shown.

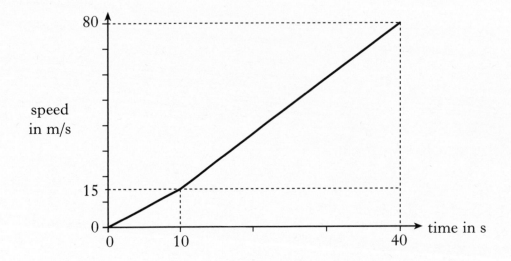

(i) Calculate the acceleration of the aircraft **during the first 10 s**.

Space for working and answer

2

(ii) Calculate the unbalanced force acting on the aircraft **during the first 10 s**.

Space for working and answer

2

Marks | K&U | PS

10. (*a*) **(continued)**

(iii) By using information **from the graph**, explain whether the unbalanced force on the aircraft is greater during the time period 0–10 s or 10 s–40 s.

....

....

....

2

(iv) Calculate the length of runway required to allow the aircraft to reach its take-off speed.

Space for working and answer

3

(*b*) After take-off, the aircraft flies at a constant height of 10 000 m. The pilot increases the speed of the aircraft at this height.

The diagram shows the forces acting on the aircraft at this height.

lift (upwards force)

engine thrust

air friction force

weight

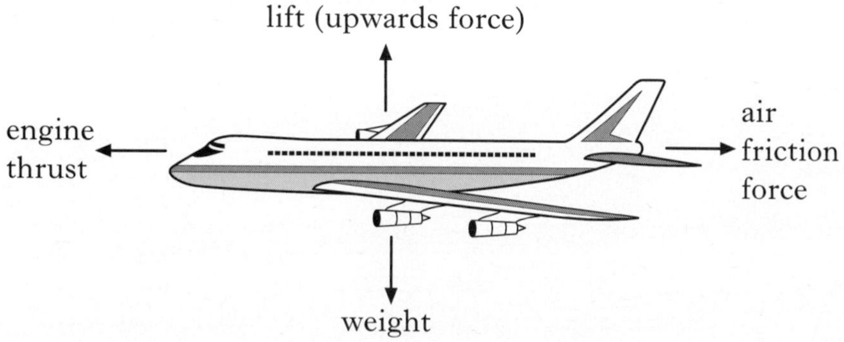

Complete the statements about the sizes of the forces acting on the aircraft by using phrases from the following list.

equal to   **greater than**   **less than**

(i) The engine thrust is................................. the air friction force.

1

(ii) The lift is...................................... the weight.

1

Marks | K&U | PS

11. (*a*) The following information relates to two power stations, a fossil fuel power station and a nuclear power station.

Fossil Fuel Power Station	Nuclear Power Station
Heat energy produced per kilogram of fuel $4 \cdot 5 \times 10^7 \text{J}$	Heat energy produced per kilogram of fuel $4 \cdot 4 \times 10^{11} \text{J}$
Waste produced per year —not radioactive 100 000 kg	Waste produced per year —radioactive 5 kg
Cooling water required 550 kg/s	Cooling water required 550 kg/s

(i) Compare the information given for the two types of power station. State **one** advantage of generating electricity using each type of power station.

Fossil fuel ....

....

Nuclear..

.... **2**

(ii) Using information given, state where both types of power station are likely to be located.

Explain why they are built in these locations.

....

....

.... **2**

(*b*) A simple block diagram of a nuclear power station is shown below.

Reactor core	→	Turbine	→	Generator

State the energy transformation that takes place in

(i) the reactor core

.... **1**

(ii) the generator.

.... **1**

Marks | K&U | PS

11. (continued)

(*c*) The diagram shows what happens when a uranium nucleus undergoes fission in a nuclear reaction.

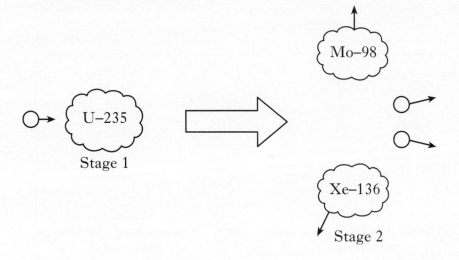

Mo–98

U–235

Stage 1

Xe–136

Stage 2

(i) Circle **one** word in each set of brackets to describe what happens at each stage.

Stage 1: A uranium nucleus is bombarded by a $\begin{pmatrix} \text{proton} \\ \text{neutron} \\ \text{electron} \end{pmatrix}$.

Stage 2: The uranium nucleus disintegrates, producing fission

fragments, two $\begin{pmatrix} \text{protons} \\ \text{neutrons} \\ \text{electrons} \end{pmatrix}$ and $\begin{pmatrix} \text{plutonium} \\ \text{heat} \\ \text{electricity} \end{pmatrix}$.

3

(ii) Describe how, in a nuclear reactor, the above process can result in a chain reaction.

....

....

.... 3

[Turn over

Marks | K&U | PS

12. Ice cubes are used to cool down water for drinking. Each ice cube has a mass of 12 g and is initially at a temperature of 0 °C.

(a) Calculate how much heat is needed to melt an ice cube.

(Any additional information needed can be found in the data sheet on page 2.)

> *Space for working and answer*

3

(b) When an ice cube is added to water, where does most of the energy come from to melt the ice?

.... **1**

(c) (i) An ice cube is added to a glass containing 200 g of water.

The initial temperature of the water is 18 °C. The final temperature when all of the ice has melted is 15 °C.

Calculate the heat removed from the water.

(Any additional information needed can be found in the data sheet on page 2.)

> *Space for working and answer*

3

(ii) Suggest a final temperature when an ice cube is added to an insulated bottle of water. The bottle has a lid and contains an equal mass of water as above, and is at the same initial temperature.

Explain your answer.

....

....

.... **2**

Marks | K&U | PS

13. Read the following passage about the launching of a space observatory using the Space Shuttle Columbia.

In July 1999, NASA used the Space Shuttle Columbia to launch a space-based observatory, called the Chandra X-ray Observatory.

This observatory is designed to detect X-rays emitted by objects in our solar system and beyond. X-rays are absorbed by the Earth's atmosphere, so a space-based observatory is necessary to detect them. Signals are sent from the observatory to Earth using radio waves.

There are now three observatories orbiting the Earth. The other two are the Hubble Space Telescope that detects visible light and the Compton Gamma Ray Observatory.

(*a*) Why is it necessary to site an observatory in space to detect X-rays?

....

.... **1**

(*b*) Four members of the electromagnetic spectrum are mentioned in the passage. Complete the diagram by placing these members in the correct order of wavelength.

		Ultraviolet		Infrared	Microwaves	

The electromagnetic spectrum **4**

(*c*) Explain why different kinds of observatory are used to detect signals from space.

....

....

.... **2**

[Turn over

Marks | K&U | PS

13. (continued)

(*d*) When the Space Shuttle reached the correct height above Earth, the observatory was separated from it.

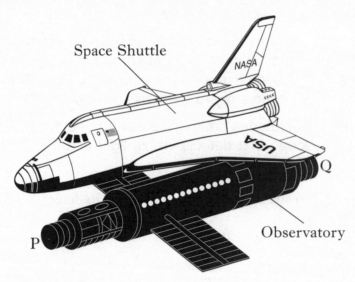

Two rocket motors P and Q on the observatory, as shown, were used during the separation. The observatory accelerated away from the space shuttle for a short time. It then remained at a fixed distance ahead of the space shuttle. Describe how the rockets P and Q were used during this separation.

....

....

....

.... 2

[END OF QUESTION PAPER]

[BLANK PAGE]

FOR OFFICIAL USE

C

K & U   PS

Total Marks

3220/402

NATIONAL
QUALIFICATIONS
2002

MONDAY, 27 MAY
10.50 AM – 12.35 PM

PHYSICS
STANDARD GRADE
Credit Level

Fill in these boxes and read what is printed below.

Full name of centre

Town

Forename(s)

Surname

Date of birth
Day Month Year

Scottish candidate number

Number of seat

1 All questions should be answered.

2 The questions may be answered in any order but all answers must be written clearly and legibly in this book.

3 Write your answer where indicated by the question or in the space provided after the question.

4 If you change your mind about your answer you may score it out and rewrite it in the space provided at the end of the answer book.

5 Before leaving the examination room you must give this book to the invigilator. If you do not, you may lose all the marks for this paper.

6 Any necessary data will be found in the **data sheet** on page two.

SCOTTISH
QUALIFICATIONS
AUTHORITY

Speed of light in materials

Material	Speed in m/s
Air	$3{\cdot}0 \times 10^8$
Carbon dioxide	$3{\cdot}0 \times 10^8$
Diamond	$1{\cdot}2 \times 10^8$
Glass	$2{\cdot}0 \times 10^8$
Glycerol	$2{\cdot}1 \times 10^8$
Water	$2{\cdot}3 \times 10^8$

Speed of sound in materials

Material	Speed in m/s
Aluminium	5200
Air	340
Bone	4100
Carbon dioxide	270
Glycerol	1900
Muscle	1600
Steel	5200
Tissue	1500
Water	1500

Gravitational field strengths

	Gravitational field strength on the surface in N/kg
Earth	10
Jupiter	26
Mars	4
Mercury	4
Moon	1·6
Neptune	12
Saturn	11
Sun	270
Venus	9

Specific heat capacity of materials

Material	Specific heat capacity in J/kg °C
Alcohol	2350
Aluminium	902
Copper	386
Diamond	530
Glass	500
Glycerol	2400
Ice	2100
Lead	128
Water	4180

Specific latent heat of fusion of materials

Material	Specific latent heat of fusion in J/kg
Alcohol	$0{\cdot}99 \times 10^5$
Aluminium	$3{\cdot}95 \times 10^5$
Carbon dioxide	$1{\cdot}80 \times 10^5$
Copper	$2{\cdot}05 \times 10^5$
Glycerol	$1{\cdot}81 \times 10^5$
Lead	$0{\cdot}25 \times 10^5$
Water	$3{\cdot}34 \times 10^5$

Melting and boiling points of materials

Material	Melting point in °C	Boiling point in °C
Alcohol	−98	65
Aluminium	660	2470
Copper	1077	2567
Glycerol	18	290
Lead	328	1737
Turpentine	−10	156

Specific latent heat of vaporisation of materials

Material	Specific latent heat of vaporisation in J/kg
Alcohol	$11{\cdot}2 \times 10^5$
Carbon dioxide	$3{\cdot}77 \times 10^5$
Glycerol	$8{\cdot}30 \times 10^5$
Turpentine	$2{\cdot}90 \times 10^5$
Water	$22{\cdot}6 \times 10^5$

SI Prefixes and Multiplication Factors

Prefix	Symbol	Factor	
giga	G	1 000 000 000	$= 10^9$
mega	M	1 000 000	$= 10^6$
kilo	k	1000	$= 10^3$
milli	m	0·001	$= 10^{-3}$
micro	μ	0·000 001	$= 10^{-6}$
nano	n	0·000 000 001	$= 10^{-9}$

DO NOT WRITE IN THIS MARGIN

Marks | K&U | PS

1. Radio Alba transmits on a range of frequencies from different transmitters throughout Scotland.

transmitter P
+

Aberdeen•

Stirling
•

+
transmitter Q

On a car journey from Aberdeen to Stirling a driver listens to Radio Alba. At the start of the journey she tunes to the signal transmitted from transmitter P.

(a) Complete the following passage, using some of the words from the list below. Do not use any word more than once.

amplitude   **audio**   **carrier**
frequency   **modulation**   **radio**

The transmitter transmits a.............................signal, which consists

of an ....wave and a................................wave. The

process of combining these waves is known as   2

(b) During the journey the driver finds that the signal from transmitter P fades.

   (i) Suggest a reason why the signal fades.

   ....

     1

   (ii) To continue to listen to Radio Alba, the driver re-tunes the radio to pick up the signal from transmitter Q.

   What is the difference between the carrier wave from transmitter P and that from transmitter Q?

   ....

     1

Marks | K&U | PS

2. The table gives information about artificial satellites that orbit the Earth.

Name of satellite	Period (minutes)	Height above Earth (km)	Use
Landsat	99	705	Land mapping
ERS-1		780	Monitoring sea levels
NOAA-12	102	833	Distribution of ozone layer
Early Bird	1440	35 900	Continuous telecommunication

(a) NOAA-12 uses radio waves to transmit signals relating to the ozone layer.

   (i) What is the speed of radio waves?

     1

   (ii) Calculate the time for signals to travel from NOAA-12 to an Earth station immediately below the satellite.

   Space for working and answer

                                                                                                  2

   (iii) Signals transmitted from NOAA-12 have a frequency of 137·5 MHz.

   Calculate the wavelength of these signals.

   Space for working and answer

                                                                                                  2

DO NOT
WRITE IN
THIS
MARGIN

Marks | K&U | PS

2. **(continued)**

(b) Using information about the period of Early Bird, explain why this satellite is used for continuous telecommunication between two points on the Earth's surface.

....

....

.... 2

(c) Give an approximate value, **in minutes**, for the period of orbit of ERS-1.

.... 1

(d) Landsat monitors heat emission from the land to build up a thermographic image.

Which part of the electromagnetic spectrum is detected by Landsat?

.... 1

(e) As well as artificial satellites, there is one natural satellite that orbits the Earth. Name this natural satellite.

.... 1

[Turn over

3. A student uses the circuit below in experiments to investigate how the voltage across different components varies when the current in the components is changed.

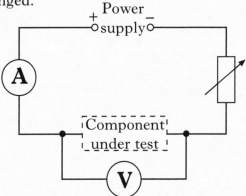

(a) The student places component X in the circuit and carries out an experiment. The graph below shows how the voltage across component X varies with current.

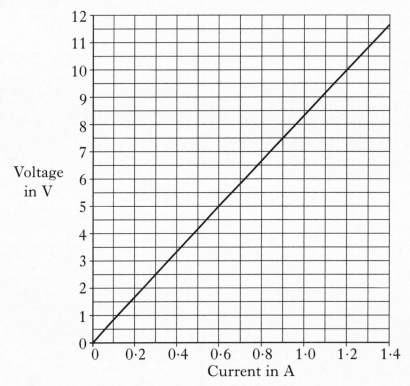

(i) Calculate the resistance of component X when the current is 1·2 A.

(You must use an appropriate number of significant figures in your answer to this question.)

Space for working and answer

2

Marks | K&U | PS

3. **(a)** **(continued)**

(ii) Using information from the graph, explain what happens to the resistance of component X as the current is increased.

Justify your answer by calculation or otherwise.

....

....

....

....

....

....

Space for working

3

(b) The student replaces component X with component Y, repeats the experiment and obtains the following graph.

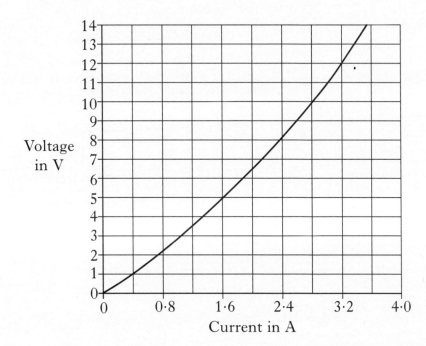

Voltage in V

Current in A

(i) The student concludes that the resistance of component Y is not constant. Why is the student correct in coming to this conclusion?

....

....

1

[Turn over

Marks | K&U | PS

3. (b) (continued)

(ii) (A) From the graph, what is the current in component Y when the voltage across component Y is 12 V?

.... 1

(B) Calculate the power dissipated in component Y when the voltage across it is 12 V.

> *Space for working and answer*

2

4. The diagram shows three household circuits, connected to a consumer unit.

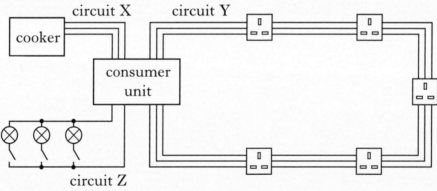

(a) (i) Which circuit is a ring circuit?

.... 1

(ii) Give **two** advantages of using a ring circuit.

....

.... 2

Marks | K&U | PS

4. (continued)

(b) State and explain **one** difference between a lighting circuit and a ring circuit.

....

....

.... **2**

(c) (i) Why does a cooker need a separate circuit?

....

....

.... **1**

(ii) One heating element of the cooker has a power rating of 2·2 kW. Calculate how many joules of energy are transferred by this element in 2 hours.

Space for working and answer

2

(d) (i) What is the purpose of an earth wire?

....

.... **1**

(ii) Explain how an earth wire works.

....

....

.... **2**

[Turn over

Marks | K&U | PS

5. Ultrasound is used by doctors for treatment and diagnosis.

(a) Pulses of ultrasound are used to produce local heating of muscle deep inside the body. This heating effect can help relieve pain in the muscles.

(i) What is meant by ultrasound?

....

.... **1**

(ii) Calculate the time for a pulse of ultrasound to travel through 2 cm of muscle.

(Data you require will be found in the Data Sheet on *page two*.)

Space for working and answer

3

(b) Ultrasound is also used to build up images of an unborn baby.

(i) Explain how ultrasound is used to build up such images.

....

....

....

.... **2**

(ii) Why is ultrasound safer than X-rays for this sort of medical application?

....

.... **1**

Official SQA Past Papers: Credit Physics 2002

DO NOT
WRITE IN
THIS
MARGIN

Marks | K&U | PS

6. A student investigates the effect of glass shapes on rays of light.

(*a*) The student places glass shapes in the path of three rays of red light as shown.

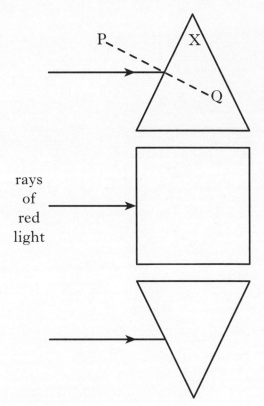

rays
of
red
light

(i) Complete the diagram to show the paths of the rays of light through and out of the three glass shapes.   3

(ii) The student has drawn line PQ on the diagram at shape X at right angles to the glass surface.

What name is given to this line?

....   1

(iii) **On the diagram**, label **one** angle of incidence as *i* and **one** angle of refraction as *r*.   2

(*b*) Name the type of lens that would have a similar effect on the rays of light as the three glass shapes, arranged as in part (*a*).

....   1

[Turn over

Marks | K&U | PS

7. The exit of an underground car park has an automatic barrier. The barrier rises when a car interrupts a light beam across the exit and money has been put into the pay machine. The barrier can also be operated by using a manual switch.

The light beam is directed at an LDR that is connected as shown in the circuit below.

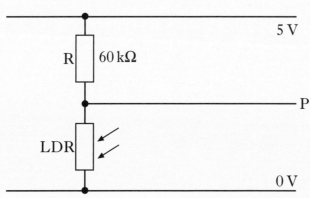

(a) Calculate the voltage across the LDR when its resistance is 15 kΩ.

> *Space for working and answer*

2

DO NOT WRITE IN THIS MARGIN

Marks | K&U | PS

7. (continued)

(*b*) Part of the control circuit for the automatic barrier is shown below.

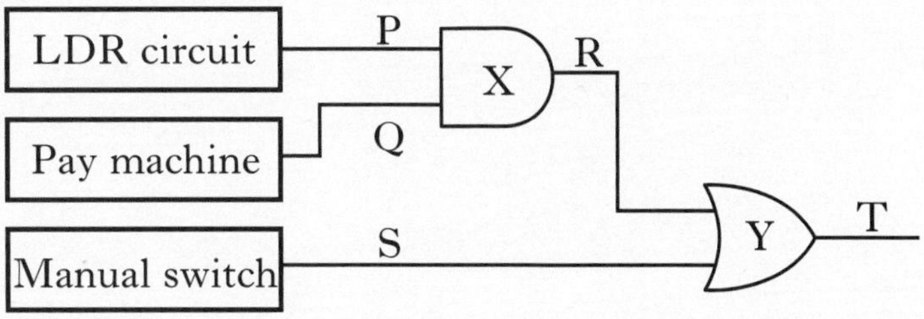

When a car interrupts the light beam, the logic level at P changes from logic 0 to logic 1.

When money is put into the pay machine, the logic level at Q changes from logic 0 to logic 1.

When the manual switch is operated, the logic level at S changes from logic 0 to logic 1.

(i) Name logic gate X.

.... 1

(ii) Name logic gate Y.

.... 1

(iii) Complete the truth table below for the control circuit shown, by filling in the values of the logic levels at R and T.

P	Q	R	S	T
0	0		0	
0	1		0	
1	0		0	
1	1		0	
0	0		1	
0	1		1	
1	0		1	
1	1		1	

4

(iv) Describe a situation where it would be necessary to operate the barrier by using the manual switch.

....

.... 1

[Turn over

DO NOT
WRITE IN
THIS
MARGIN

Marks | K&U | PS

8. A radio has three types of output device.

filament lamp   **LED**   **loudspeaker**

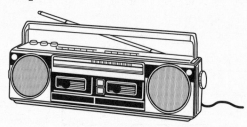

(a) Which of these output devices transforms electrical energy into sound energy?

.... 1

(b) Which of these output devices is most suitable for illuminating the front panel of the radio?

Explain your choice.

....

....

.... 2

(c) The LED is connected in series with resistor, R, to the 9·0 V power supply of the radio.

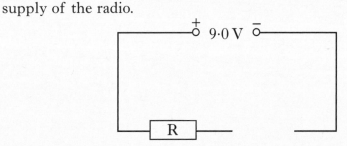

(i) In the space in the circuit above draw the LED connected correctly. 2

(ii) When lit, the voltage across the LED is 2·4 V and the current in the LED is 20 mA.

Calculate the resistance of R.

Space for working and answer

3

Marks | K&U | PS

9. A skateboarder is practising on a ramp. The total mass of the skateboarder and the board is 60 kg.

(a) Calculate the increase in potential energy of the skateboarder and board in moving from the ground to position P.

Space for working and answer

2

(b) The skateboarder moves along the ramp from P to R, and rises into the air above R.

  (i) At what point **on the ramp** is the kinetic energy of the skateboarder greatest?

.... 1

  (ii) The vertical speed of the skateboarder at R is 6 m/s.
      Calculate the height that the skateboarder rises to, above R.

Space for working and answer

3

  (iii) Explain why the skateboarder does not rise to the same height as P.

....

....

.... 2

Marks | K&U | PS

10. At a greyhound racing track, the greyhounds are automatically released when an artificial hare crosses the starting line.

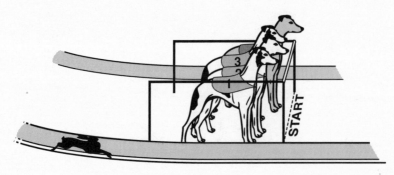

The speed-time graph shows the motion of one greyhound and the hare from the time when the hare crosses the starting line.

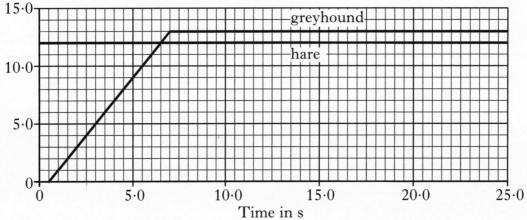

(a) How long does it take for the greyhound to start moving after the hare crosses the starting line?

.... **1**

(b) Calculate the acceleration of the greyhound when it starts moving.

Space for working and answer

2

DO NOT WRITE IN THIS MARGIN

Marks | K&U | PS

10. (continued)

(*c*) The hare crosses the finishing line 20 s after crossing the starting line.

(i) Over what distance is the race run?

Space for working and answer

2

(ii) How far behind the hare is the greyhound when the **hare** finishes the race?

Space for working and answer

3

[Turn over

DO NOT
WRITE IN
THIS
MARGIN

Marks | K&U | PS

11. A lighting system in a shop window uses three identical 18 W, 12 V filament lamps. The lamps are operated at their correct rating from the 230 V mains supply using a transformer as shown below.

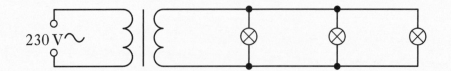

There are 5750 turns on the primary coil of the transformer.

(a) Calculate the number of turns on the secondary coil of the transformer.

> *Space for working and answer*

2

(b) (i) The current in each lamp is 1·5 A.
Calculate the total current in the secondary circuit of the transformer.

> *Space for working and answer*

1

(ii) Assuming that the transformer is 100% efficient, calculate the current in the primary coil.

> *Space for working and answer*

2

DO NOT WRITE IN THIS MARGIN

Marks | K&U | PS

11. (continued)

(*c*) (i) Show that the resistance of one of the filament lamps, when it is operating normally, is $8{\cdot}0\,\Omega$.

Space for working and answer

2

(ii) Calculate the combined resistance of the three lamps in parallel.

Space for working and answer

2

[Turn over

Marks | K&U | PS

12. A student sets up the apparatus shown to measure the specific heat capacity of an aluminium block.

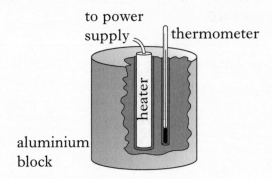

to power
supply — thermometer

heater

aluminium
block

The student obtains the following results:

mass of aluminium block $m = 0.8\,\text{kg}$
temperature change $\Delta T = 19\,°\text{C}$
time taken $t = 5.0$ minutes
heater current $I = 4.2\,\text{A}$
heater voltage $V = 12\,\text{V}$

(a) Show, by calculation, that $15\,120\,\text{J}$ of electrical energy are supplied to the heater in 5.0 minutes.

Space for working and answer

2

(b) (i) Assuming all of the electrical energy is transferred to the aluminium block as heat energy, calculate the value of the specific heat capacity of aluminium obtained from this experiment.

Space for working and answer

2

Marks | K&U | PS

12. *(b)* **(continued)**

(ii) The accepted value of the specific heat capacity of aluminium is 902 J/kg °C.

(A) Give a reason for the difference between your answer in *(b)*(i) and this value.

....

....

.... 1

(B) How could the experiment be improved to reduce this difference?

....

....

.... 1

[Turn over

Marks | K&U | PS

13. During the Apollo 11 expedition to the Moon, 21 kg of soil samples were brought from the Moon to the Earth. The gravitational field strength was not constant throughout the journey.

(a) What is meant by gravitational field strength?

....

....   1

(b) Complete the table to show the mass and weight of the soil samples at various stages of the journey.

Stage	Gravitational field strength (N/kg)	Mass (kg)	Weight (N)
on the Moon	1·6	21	
at a point during the journey	0		
on the Earth	10		

3

[END OF QUESTION PAPER]

[BLANK PAGE]

C

K & U   PS

Total Marks

3220/402

NATIONAL
QUALIFICATIONS
2003

MONDAY, 19 MAY
10.50 AM – 12.35 PM

PHYSICS
STANDARD GRADE
Credit Level

Fill in these boxes and read what is printed below.

Full name of centre

Town

Forename(s)

Surname

Date of birth
Day Month Year

Scottish candidate number

Number of seat

1 All questions should be answered.

2 The questions may be answered in any order but all answers must be written clearly and legibly in this book.

3 Write your answer where indicated by the question or in the space provided after the question.

4 If you change your mind about your answer you may score it out and rewrite it in the space provided at the end of the answer book.

5 Before leaving the examination room you must give this book to the invigilator. If you do not, you may lose all the marks for this paper.

6 Any necessary data will be found in the **data sheet** on page two.

SCOTTISH
QUALIFICATIONS
AUTHORITY

DATA SHEET

Speed of light in materials

Material	Speed in m/s
Air	$3 \cdot 0 \times 10^8$
Carbon dioxide	$3 \cdot 0 \times 10^8$
Diamond	$1 \cdot 2 \times 10^8$
Glass	$2 \cdot 0 \times 10^8$
Glycerol	$2 \cdot 1 \times 10^8$
Water	$2 \cdot 3 \times 10^8$

Speed of sound in materials

Material	Speed in m/s
Aluminium	5200
Air	340
Bone	4100
Carbon dioxide	270
Glycerol	1900
Muscle	1600
Steel	5200
Tissue	1500
Water	1500

Gravitational field strengths

	Gravitational field strength on the surface in N/kg
Earth	10
Jupiter	26
Mars	4
Mercury	4
Moon	$1 \cdot 6$
Neptune	12
Saturn	11
Sun	270
Venus	9

Specific heat capacity of materials

Material	Specific heat capacity in J/kg °C
Alcohol	2350
Aluminium	902
Copper	386
Diamond	530
Glass	500
Glycerol	2400
Ice	2100
Lead	128
Water	4180

Specific latent heat of fusion of materials

Material	Specific latent heat of fusion in J/kg
Alcohol	$0 \cdot 99 \times 10^5$
Aluminium	$3 \cdot 95 \times 10^5$
Carbon dioxide	$1 \cdot 80 \times 10^5$
Copper	$2 \cdot 05 \times 10^5$
Glycerol	$1 \cdot 81 \times 10^5$
Lead	$0 \cdot 25 \times 10^5$
Water	$3 \cdot 34 \times 10^5$

Melting and boiling points of materials

Material	Melting point in °C	Boiling point in °C
Alcohol	−98	65
Aluminium	660	2470
Copper	1077	2567
Glycerol	18	290
Lead	328	1737
Turpentine	−10	156

Specific latent heat of vaporisation of materials

Material	Specific latent heat of vaporisation in J/kg
Alcohol	$11 \cdot 2 \times 10^5$
Carbon dioxide	$3 \cdot 77 \times 10^5$
Glycerol	$8 \cdot 30 \times 10^5$
Turpentine	$2 \cdot 90 \times 10^5$
Water	$22 \cdot 6 \times 10^5$

SI Prefixes and Multiplication Factors

Prefix	Symbol	Factor	
giga	G	1 000 000 000	$= 10^9$
mega	M	1 000 000	$= 10^6$
kilo	k	1000	$= 10^3$
milli	m	$0 \cdot 001$	$= 10^{-3}$
micro	μ	$0 \cdot 000 001$	$= 10^{-6}$
nano	n	$0 \cdot 000 000 001$	$= 10^{-9}$

1. A farm road joins a main road at a bend. The farmer has placed a mirror as shown so that he can see when cars are approaching.

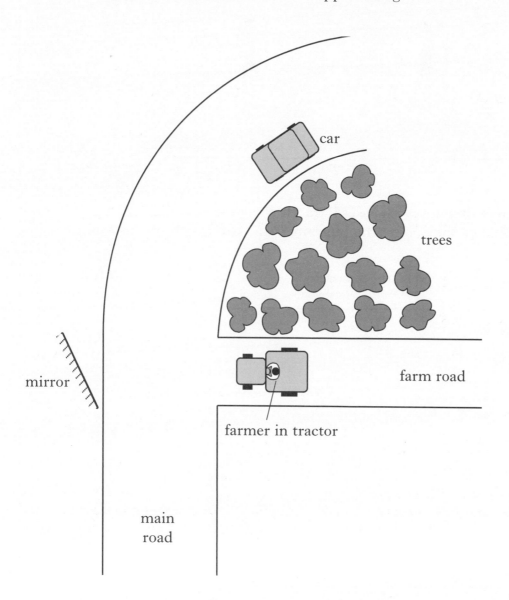

(a) On the diagram, draw rays to show how the farmer in the tractor can see the car by using the mirror.

You must label the angle of incidence and the angle of reflection on your completed diagram.

3

(b) State why the driver of the car can **also** see the tractor using the mirror.

....

....

1

Marks | K&U | P

2. Two students watch the waves produced by a wave machine at a swimming pool.

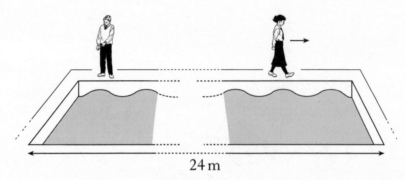

24 m

One student walks beside a wave as it travels along the pool. The wave goes from one end of the pool to the other in 20 s. The length of the pool is 24 m.

(a) Calculate the speed of the waves.

Space for working and answer

2

(b) In the same time interval, the other student counts 5 waves going past the point where he is standing.

Calculate the frequency of the waves.

Space for working and answer

2

DO NOT WRITE IN THIS MARGIN

Marks | K&U | PS

2. (continued)

(*c*) The students note that there are 5 complete waves in the pool at any time.

Calculate the wavelength of the waves.

Space for working and answer

2

(*d*) Explain why "distance divided by time" and "frequency times wavelength" are equivalent for a wave.

Space for working and answer

2

[Turn over

Marks

3. A home entertainment centre consists of four appliances. The table gives the power rating of each appliance.

Appliance	Power rating (W)
television	110
video recorder	22
satellite receiver	20
DVD player	18

To operate properly, each appliance must be connected to mains voltage. The appliances are connected to the mains using a multiway adaptor.

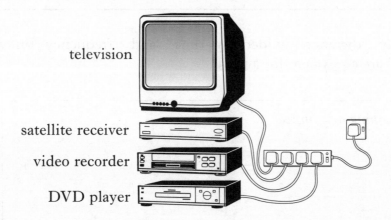

television

satellite receiver

video recorder

DVD player

(a) (i) State the value of the operating voltage of the appliances.

.... 1

(ii) The connections in the multiway adaptor are arranged to ensure that each appliance is connected to mains voltage.

State how the connections in the multiway adaptor are arranged to achieve this.

.... 1

(b) Calculate the current from the mains when all four appliances are operating at the power ratings shown in the table.

(You must use an appropriate number of significant figures in your answer to this question.)

Space for working and answer

3

Marks | K&U | PS

3. **(continued)**

(*c*) Calculate the resistance of the television when it is operating at the power rating stated in the table.

> *Space for working and answer*

2

(*d*) The plug on the flex of the multiway adaptor contains a fuse.

What is the purpose of this fuse?

....

....

.... 1

[Turn over

Page seven

Marks | K&U | PS

4. A show uses five spotlights of equal brightness, pointing at the same place on the stage.

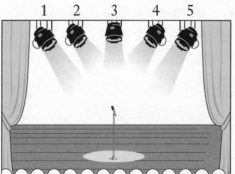

The spotlights can be turned on and off individually. The colour of light from each spotlight is shown in the table.

Spotlight	Colour
1	green
2	blue
3	red
4	blue
5	green

(a) State **three** spotlights that could be on to produce white light on the stage.

....

.... 1

(b) One scene requires yellow light.

State **two** spotlights that could be on to produce yellow light on the stage.

....

.... 1

(c) Another scene requires **pale** green light. This needs **four** of the spotlights to be on.

State **one** spotlight that could be **off** so that the other four produce pale green light.

.... 1

Marks | K&U | PS

5. A textbook has three diagrams showing how an eye lens changes when looking at objects that are different distances away. The diagrams below are copies of these three diagrams, with parts omitted.

Diagrams 1 and 3 are not complete.

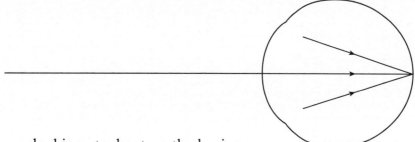

Diagram 1   looking at a boat on the horizon

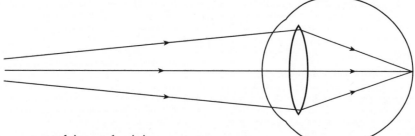

Diagram 2   watching television across a room

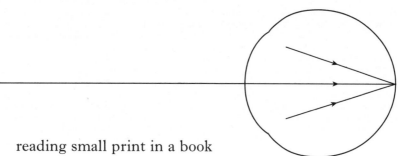

Diagram 3   reading small print in a book

(a) On diagrams 1 **and** 3:

(i)   draw two rays to show light coming from each object to the eye;

(ii) draw a lens to show how the shape of the eye lens is different from the shape of the lens in diagram 2.   **4**

(b) The focal length of an eye lens system (the cornea and the eye lens together) is 2·5 cm.

Calculate the power of this eye lens system.

Space for working and answer

2

Marks | K&U | PS

6. A student designs the circuit shown to operate a 12 V, 3 A lamp from a 36 V supply.

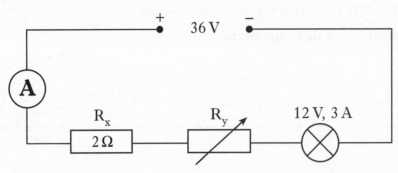

(a) What is the reading on the ammeter when the lamp is operating at its correct power rating?

.... **1**

(b) The resistance of R_x is $2\,\Omega$.

Calculate the voltage across R_x when the lamp is operating correctly.

Space for working and answer

2

(c) Calculate the resistance of R_y when the lamp is operating correctly.

Space for working and answer

3

Marks | K&U | PS

6. (continued)

(*d*) The student connects a second, identical lamp as shown in the diagram below.

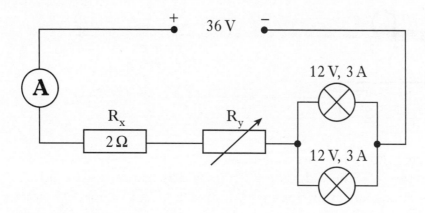

Explain why the resistance of R_y has to be adjusted for both lamps to operate correctly.

....

....

.... 2

[Turn over

Page eleven

Marks | K&U | PS

7. A paper mill uses a radioactive source in a system to monitor the thickness of paper.

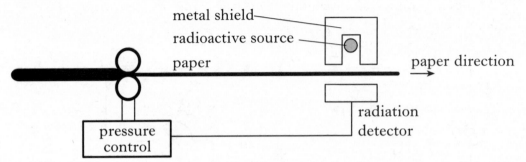

The count rate detected by the radiation detector changes as the thickness of the paper varies. The radiation detector sends signals to the pressure control to maintain an even thickness of paper. The radioactive source emits a type of radiation that is partly absorbed by the paper. The source also has a half-life that allows the mill to run continuously, for several days.

(a) What is meant by the term "half-life"?

....

.... 1

(b) The following radioactive sources are available.

Source	Half-life	Radiation emitted
P	500 years	alpha
Q	20 hours	beta
R	450 years	beta
S	300 years	gamma

(i) Explain why source P cannot be used in this system.

....

....

.... 1

(ii) Which source should be used? Explain your answer.

....

....

.... 2

Marks | K&U | PS

7. **(continued)**

(*c*) Why does the radioactive source in the paper mill have a metal shield?

....

....   1

(*d*) Another radioactive source emits gamma radiation. The graph shows how the activity of this source decreases with time.

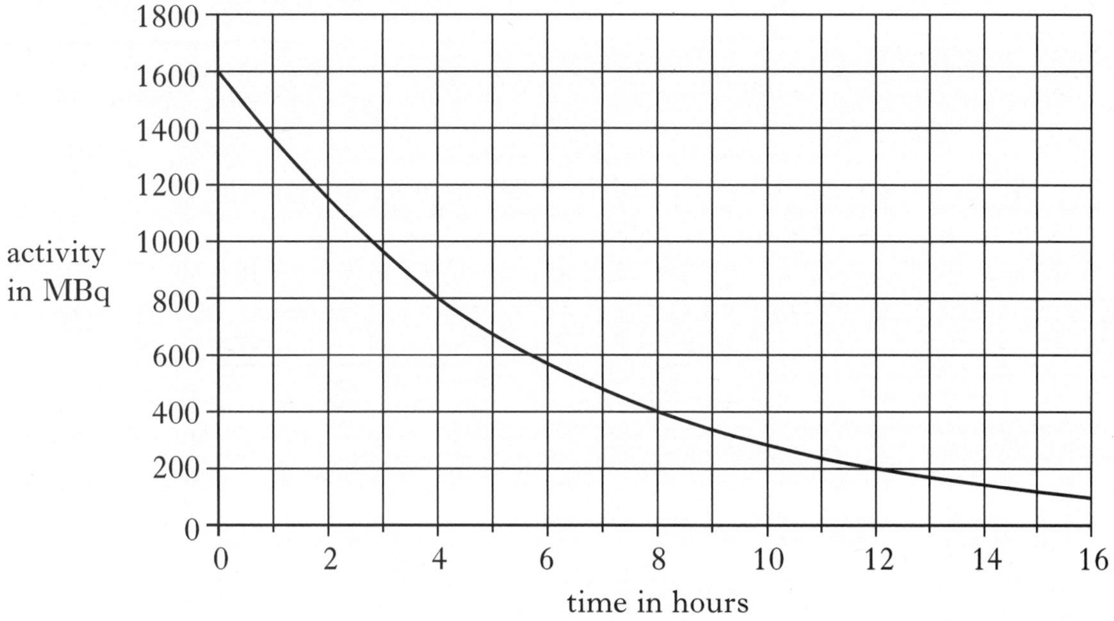

activity in MBq

time in hours

Calculate the half-life of this radioactive source.

Space for working and answer

1

[Turn over

Marks | K&U | PS

8. A bus is fitted with a buzzer that sounds only when the bus is reversing. Part of the circuit that operates the buzzer is shown.

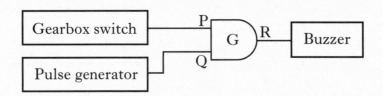

The output from the gearbox switch is high (logic 1) when the bus is reversing.

(a) Name logic gate G.

.... 1

(b) The table shows the different possible combinations of logic levels (0 or 1) for input P and input Q to gate G.

Complete the last column of the table by **drawing** the output R from gate G for each combination of inputs.

Input P	Input Q	Output R
1	1	1
0 ———	0 ———	0
1	1 ⎍⎍	1
0 ———	0	0
1 ———	1	1
0	0 ———	0
1 ———	1 ⎍⎍	1
0	0	0

2

(c) The pulse generator part of the circuit is shown below.

The power supply to the NOT gate has been omitted for clarity.

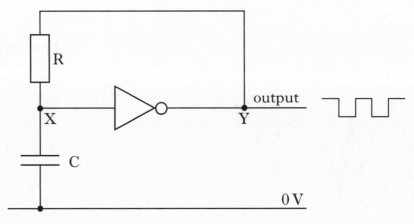

[3220/402]

Marks | K&U | PS

8. **(*c*)** **(continued)**

(i) Capacitor C is initially discharged.

Explain the operation of the pulse generator circuit, by referring to points X and Y in the circuit.

....

....

....

....

.... **2**

(ii) The pulse generator produces an output with a high frequency.

State **one** change that could be made to the circuit to give an output of lower frequency.

....

.... **1**

[Turn over

9. An electronic circuit is shown below. Component R is a thermistor.

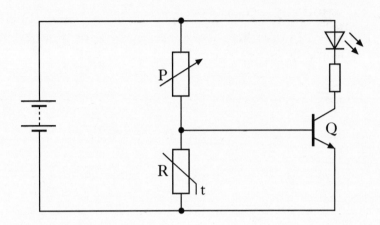

(*a*) Name component P.

.... 1

(*b*) (i) Name component Q.

.... 1

(ii) **In this circuit**, what is the function of component Q?

.... 1

(*c*) Explain how the circuit operates.

....

....

....

....

....

.... 2

Marks | K&U | PS

10. A cyclist starts a journey in first gear and uses two other gears during the journey. After a short time the cyclist is forced to brake sharply and comes to a halt. A speed-time graph of the journey is shown.

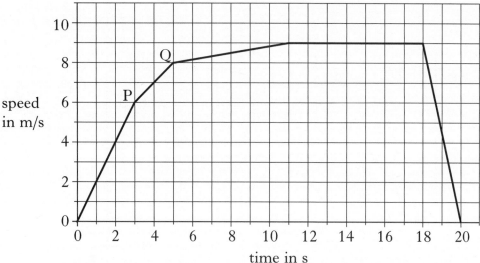

At point P the cyclist changes from first gear to second gear.
At point Q the cyclist changes from second gear to third gear.

(*a*) (i) Before braking, which gear is the cyclist using when the acceleration is greatest?

    .... **1**

   (ii) Which gear does the cyclist use for the shortest time?

    .... **1**

(*b*) Calculate how far the cyclist travels in second gear.

Space for working and answer

3

(*c*) Calculate the deceleration.

Space for working and answer

2

Page seventeen   **[Turn over**

Marks | K&U | PS

11. A model motor boat of mass 4 kg is initially at rest on a pond. The boat's motor, which provides a constant force of 5 N, is switched on. As the boat accelerates, the force of friction acting on it increases. A graph of the force of friction acting on the boat against time is shown.

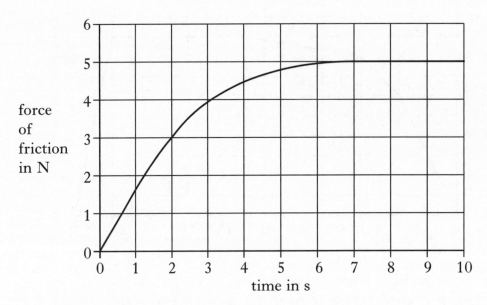

(a) (i) State the force of friction acting on the boat 2 s after the motor is switched on.

.... 1

(ii) Calculate the acceleration of the boat at this time.

Space for working and answer

3

(b) Describe and explain the movement of the boat after 7 s.

....

....

.... 2

12. A battery charger with an input voltage of 230 V is used to recharge a car battery. The charger contains a transformer that has an output voltage of 13·8 V.

(a) What type of transformer does the battery charger contain?

.... **1**

(b) There are 4000 turns in the primary coil of the transformer.

Assuming the transformer is 100% efficient, calculate the number of turns in the secondary coil.

Space for working and answer

2

(c) (i) When charging the battery, the current in the secondary coil is 4·7 A.

(A) Calculate the power output of the transformer.

Space for working and answer

2

(B) In practice, the transformer is only 94% efficient.
Calculate the current in the primary coil.

Space for working and answer

3

(ii) State and explain **one** reason why a transformer is not 100% efficient.

....

....

.... **2**

Marks | K&U | PS

13. Water from a stream is used to drive a water wheel. The stream provides 6000 kg of water per minute to the wheel. The water falls a vertical height of 5 m.

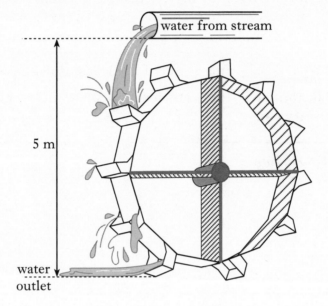

water from stream

5 m

water outlet

(a) Show that the maximum power available to the wheel from the water is 5000 W.

Space for working and answer

3

(b) The water wheel turns an electrical generator. The generator produces an output of 2990 W.

  (i) Calculate the efficiency of the water wheel and generator system.

Space for working and answer

2

Marks | K&U | PS

13. **(b)** **(continued)**

(ii) Give **two** reasons why the efficiency of this system is not 100%.

....

....

....

....

2

(iii) The generator is connected to a heater in a shed. The heater heats the air in the shed. The mass of air in the shed is 161 kg. The specific heat capacity of air is 1000 J/kg °C.

Calculate the minimum time to increase the temperature of the air in the shed by 13 °C.

Space for working and answer

3

(iv) Give **one** reason why the actual time taken to increase the temperature of the air in the shed is greater than the value calculated in (iii).

....

....

1

[Turn over

DO NOT
WRITE IN
THIS
MARGIN

Marks | K&U | PS

14. Gamma rays, ultraviolet and infrared are three members of a family of waves. Every member of this family travels at the speed of light.

(a) What name is given to this family of waves?

.... 1

(b) Some uses of waves in this family are shown below.

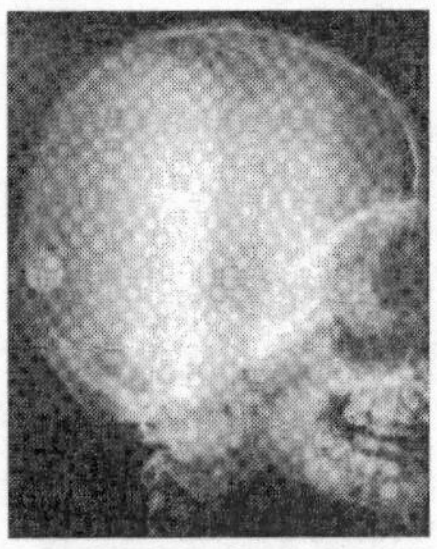

Photographing
bones inside a
body

Tanning with a
sun-ray lamp

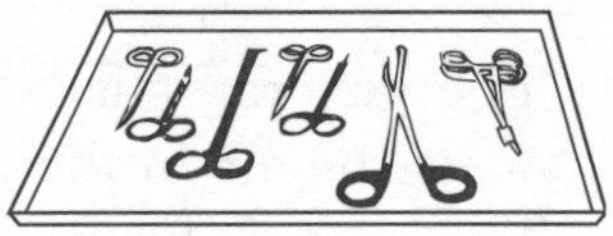

Sterilising medical
instruments

Communicating
with mobile
phones

Linking networked
computers through
optical fibres

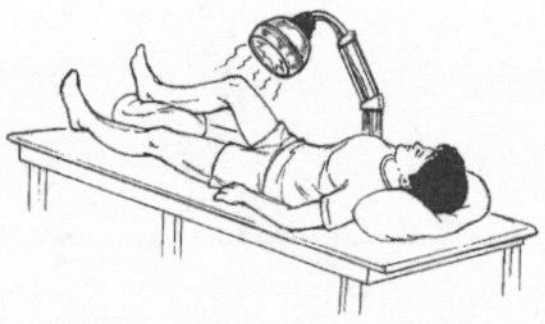

Treating injuries
using a heat-lamp

(i) From the examples above, give a use for:

gamma rays..

ultraviolet ....

infrared.. 3

(ii) Which of the three waves in (i) has:

the longest wavelength ....

the highest frequency? 2

15. A darts player throws a dart horizontally at the centre of the inner bull. The dart leaves the player's hand at a distance of 2·16 m from the dartboard and with a horizontal speed of 12·0 m/s.

inner bull

2·16 m

(a) Calculate the time taken for the dart to travel from the hand to the board.

Space for working and answer

2

(b) Explain why the dart follows a curved path in its flight to the board.

....

....

....

2

(c) The average vertical speed of the dart during its flight to the board is 0·9 m/s.

How far below the centre of the inner bull does the dart hit the board?

Space for working and answer

2

[END OF QUESTION PAPER]

DO NOT WRITE IN THIS MARGIN

K&U | PS

YOU MAY USE THE SPACE ON THIS PAGE TO REWRITE ANY ANSWER YOU HAVE DECIDED TO CHANGE IN THE MAIN PART OF THE ANSWER BOOKLET. TAKE CARE TO WRITE IN CAREFULLY THE APPROPRIATE QUESTION NUMBER.

[BLANK PAGE]

FOR OFFICIAL USE

C

K & U   PS

Total Marks

3220/402

NATIONAL
QUALIFICATIONS
2004

FRIDAY, 28 MAY
10.50 AM – 12.35 PM

PHYSICS
STANDARD GRADE
Credit Level

Fill in these boxes and read what is printed below.

Full name of centre

Town

Forename(s)

Surname

Date of birth
Day Month Year   Scottish candidate number   Number of seat

1 All questions should be answered.

2 The questions may be answered in any order but all answers must be written clearly and legibly in this book.

3 Write your answer where indicated by the question or in the space provided after the question.

4 If you change your mind about your answer you may score it out and rewrite it in the space provided at the end of the answer book.

5 Before leaving the examination room you must give this book to the invigilator. If you do not, you may lose all the marks for this paper.

6 Any necessary data will be found in the **data sheet** on page two.

7 Care should be taken to give an appropriate number of significant figures in the final answers to questions.

SCOTTISH
QUALIFICATIONS
AUTHORITY

DATA SHEET

Speed of light in materials

Material	Speed in m/s
Air	$3{\cdot}0 \times 10^8$
Carbon dioxide	$3{\cdot}0 \times 10^8$
Diamond	$1{\cdot}2 \times 10^8$
Glass	$2{\cdot}0 \times 10^8$
Glycerol	$2{\cdot}1 \times 10^8$
Water	$2{\cdot}3 \times 10^8$

Speed of sound in materials

Material	Speed in m/s
Aluminium	5200
Air	340
Bone	4100
Carbon dioxide	270
Glycerol	1900
Muscle	1600
Steel	5200
Tissue	1500
Water	1500

Gravitational field strengths

	Gravitational field strength on the surface in N/kg
Earth	10
Jupiter	26
Mars	4
Mercury	4
Moon	1·6
Neptune	12
Saturn	11
Sun	270
Venus	9

Specific heat capacity of materials

Material	Specific heat capacity in J/kg °C
Alcohol	2350
Aluminium	902
Copper	386
Diamond	530
Glass	500
Glycerol	2400
Ice	2100
Lead	128
Water	4180

Specific latent heat of fusion of materials

Material	Specific latent heat of fusion in J/kg
Alcohol	$0{\cdot}99 \times 10^5$
Aluminium	$3{\cdot}95 \times 10^5$
Carbon dioxide	$1{\cdot}80 \times 10^5$
Copper	$2{\cdot}05 \times 10^5$
Glycerol	$1{\cdot}81 \times 10^5$
Lead	$0{\cdot}25 \times 10^5$
Water	$3{\cdot}34 \times 10^5$

Melting and boiling points of materials

Material	Melting point in °C	Boiling point in °C
Alcohol	−98	65
Aluminium	660	2470
Copper	1077	2567
Glycerol	18	290
Lead	328	1737
Turpentine	−10	156

Specific latent heat of vaporisation of materials

Material	Specific latent heat of vaporisation in J/kg
Alcohol	$11{\cdot}2 \times 10^5$
Carbon dioxide	$3{\cdot}77 \times 10^5$
Glycerol	$8{\cdot}30 \times 10^5$
Turpentine	$2{\cdot}90 \times 10^5$
Water	$22{\cdot}6 \times 10^5$

SI Prefixes and Multiplication Factors

Prefix	Symbol	Factor	
giga	G	1 000 000 000	$= 10^9$
mega	M	1 000 000	$= 10^6$
kilo	k	1000	$= 10^3$
milli	m	0·001	$= 10^{-3}$
micro	μ	0·000 001	$= 10^{-6}$
nano	n	0·000 000 001	$= 10^{-9}$

Marks | K&U | PS

1. A mobile phone can send signals on 3 different frequencies, 900 MHz, 1800 MHz and 1900 MHz.

   (a) (i) Which signal has the longest wavelength?

   1

   (ii) Calculate the wavelength of the 1800 MHz signal.

   > *Space for working and answer*

   3

   (b) At a base station, microwave signals from the mobile phone are converted into light signals for transmission along an optical fibre.

   (i) State two advantages of sending light signals along an optical fibre compared to sending electrical signals along a wire.

   ....

   2

   (ii) The time taken for light to travel along a glass optical fibre is 1·2 ms.

   (A) State the speed at which signals travel along the optical fibre.

   1

   (B) Calculate the length of the optical fibre.

   > *Space for working and answer*

   2

DO NOT WRITE IN THIS MARGIN

Marks | K&U | PS

2. A colour television receiver displays 25 images on the screen every second.

(*a*) Calculate the number of images displayed on the screen in one minute.

Space for working and answer

1

(*b*) The television receiver contains decoders.

State the function of a decoder.

....

.... **1**

(*c*) In the colour television tube, three electron guns each send a beam of electrons to the screen.

(i) Why are **three** electron guns needed in a **colour** television tube?

....

.... **1**

(ii) The diagram below shows the screen and the shadow mask in a colour television tube.

screen
red
green
blue
R
G
B
metal shadow mask
3 electron beams from 3 electron guns

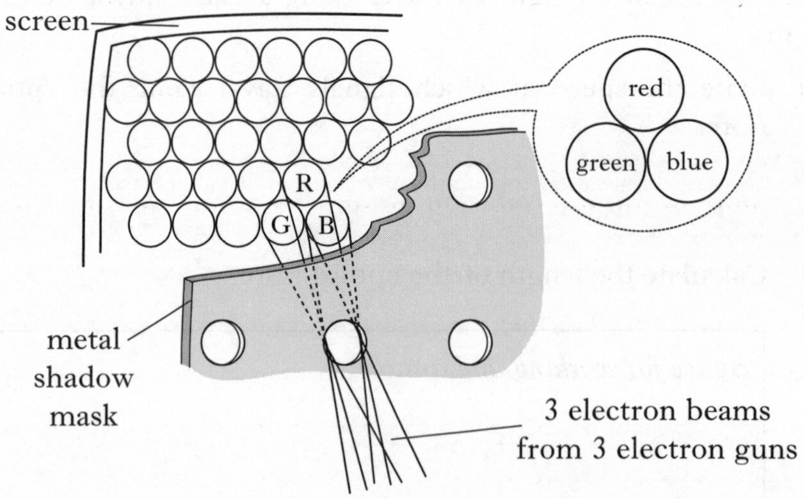

Use information from the diagram to explain why a shadow mask is needed.

....

....

.... **2**

3. A portable radio contains a rechargeable battery and a generator. The battery is charged by turning the handle of the generator.

(a) State the purpose of the battery.

.... **1**

(b) The battery is fully discharged. The handle of the generator is turned 500 times by a constant force of 9·0 N. For each turn of the handle, the force moves through a distance of 400 mm.

   (i) Show that the work done in charging the battery is 1800 J.

          Space for working and answer

          2

   (ii) Only 90% of the work done in charging the battery is available as output energy from the battery.

     (A) Calculate the output energy available.

           Space for working and answer

           2

     (B) When operating, the radio takes a current of 250 mA. The voltage of the battery is 3 V.

       Calculate the maximum time for which the radio operates.

           Space for working and answer

           2

Marks | K&U | PS

4. The circuit diagram of the wiring of a car's sidelights and headlights is shown.

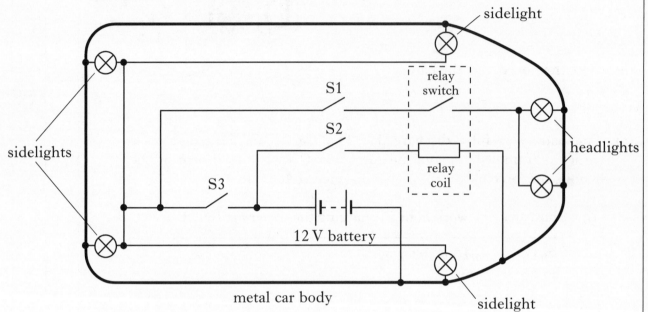

S1 is the headlight switch. S2 is the ignition switch.

When there is a current in the relay coil, the relay switch closes.

(a) Which lights are on when switch S3 **only** is closed?

.... 1

(b) At night the car has the sidelights on and the headlights on. The driver switches off the ignition. This opens the ignition switch.

Explain why **only** the headlights go out.

....

....

.... 2

4. (continued)

(*c*) **Each** sidelight is rated at 12 V, 6 W, and **each** headlight is rated at 12 V, 55 W.

(i) Calculate the current in the battery when **only** the sidelights are on.

> *Space for working and answer*

3

(ii) The driver leaves the car for 10 minutes with **only** the sidelights on.

Calculate the charge that flows through the battery in this time.

> *Space for working and answer*

2

(iii) Each headlight gives out more light energy than each sidelight when on for the same time.

Explain why this happens.

....

....

.... **2**

[Turn over

DO NOT
WRITE IN
THIS
MARGIN

Marks | K&U | PS

5. An entry system for a block of flats lets residents speak to callers before unlocking the outside door.

(a) A microphone at the outside door is connected through an amplifier to a loudspeaker in a flat.

microphone 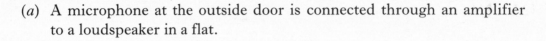 amplifier 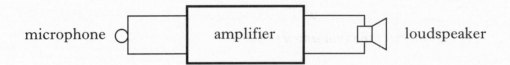 loudspeaker

The input power to the amplifier from the microphone is 5 mW and the output power from the amplifier is 2 W.

(i) Calculate the power gain of the amplifier.

Space for working and answer

2

(ii) The voltage across the loudspeaker is 4 V.

Calculate the resistance (impedance) of the loudspeaker.

Space for working and answer

2

Official SQA Past Papers: Credit Physics 2004

DO NOT
WRITE IN
THIS
MARGIN

Marks | K&U | PS

5. (continued)

(b) The entry system allows a resident to unlock the outside door from the flat. The diagram below shows this part of this system.

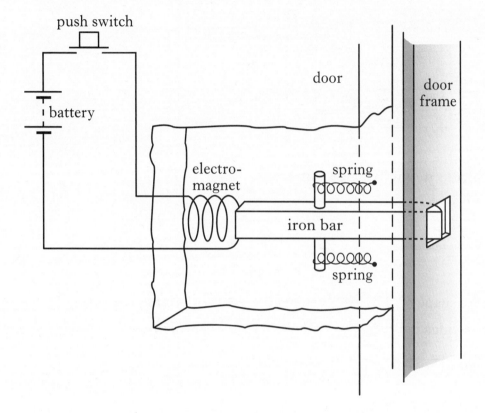

Explain how this part of the system operates to unlock the door.

....

....

....

.... 2

[Turn over

Marks | K&U | PS

6. A person visits an optician for an eye test and is found to be long sighted in both eyes. The optician issues the following prescription for lenses.

	Power of lens required (D)
Left eye	+2·5
Right eye	+1·0

(a) State what is meant by long sight.

....

....   1

(b) Draw the shape of the lenses used to correct the defect in each eye.

Your drawings must show how the two lenses are different.

Shape of lens for left eye	
Shape of lens for right eye	

3

(c) Calculate the focal length of the lens prescribed for the left eye.

Space for working and answer

2

Marks | K&U | PS

7. A smoke detector contains two metal electrodes, a battery and an alarm circuit. Alpha radiation from a radioactive source ionises air between the two electrodes.

A voltage is applied across the electrodes. Although there is a gap between the two electrodes, there is a current between the electrodes. When there are smoke particles between the electrodes, this current is reduced. This sets off the alarm.

(*a*) (i) What is meant by ionisation?

....

.... **1**

    (ii) Explain how the current is produced in the gap between the electrodes.

....

.... **1**

(*b*) Apart from safety reasons, why is a source that emits alpha radiation more suitable in a smoke detector than a source that emits gamma radiation?

....

.... **1**

(*c*) State the unit of activity of a radioactive source.

.... **1**

8. At a bottling plant, shampoo bottles on a conveyor pass a liquid level detector. Bottles filled to an acceptable level continue along the conveyor for packing. Bottles that are overfilled or underfilled are rejected.

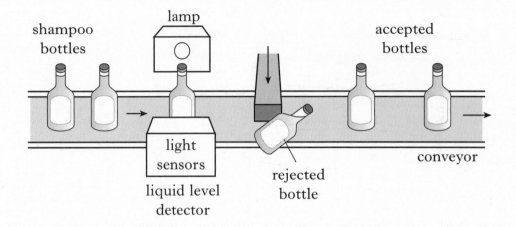

The liquid level detector consists of a lamp and two identical light sensors. The sensors are placed as shown in the diagram below. Light from the lamp can reach a sensor only when there is no shampoo between the lamp and the sensor.

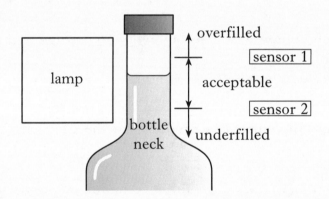

Part of the logic circuit of the liquid level detector is shown below.

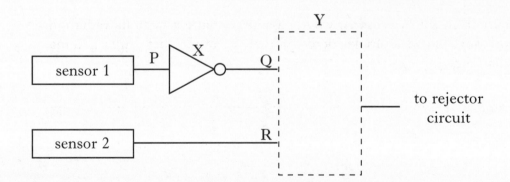

8. (continued)

The logic level outputs of a light sensor are as shown.

Light level at sensor	Logic level output
dark	0
light	1

(a) Name gate X.

.... **1**

(b) Complete the table to show the logic levels at P, Q and R when bottles filled to different levels are at the detector.

Liquid level	P	Q	R
Overfilled			
Acceptable			
Underfilled			

3

(c) The rejector circuit requires a logic level 1 to operate.

What type of gate at Y gives a logic 1 output only when a bottle is not filled to an acceptable level?

.... **1**

[Turn over

Marks | K&U | PS

9. Land speed records are calculated by timing a vehicle as it travels a measured distance of 2·0 km.

   (a) Explain whether the average speed or the instantaneous speed of the vehicle can be calculated from these measurements.

   ....

   ....

   .... **2**

   (b) A vehicle travels the measured distance at a constant speed of 220 m/s. Calculate the time taken.

   > *Space for working and answer*

   2

   (c) At the end of the measured distance, the driver switches off the engine and opens a parachute to brake.

   The speed-time graph shows the motion of the vehicle from this time.

   The mass of the vehicle is 3000 kg.

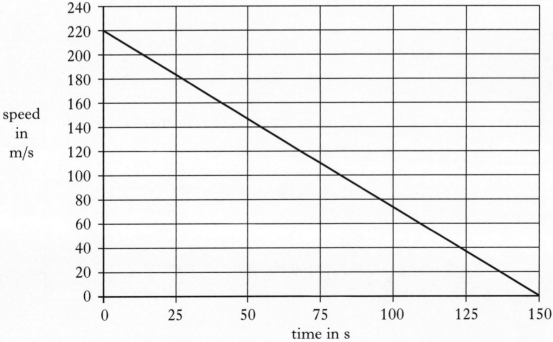

Marks | K&U | PS

9. **(c)** **(continued)**

(i) Explain how the parachute helps to reduce the speed of the vehicle.

....

.... 1

(ii) Calculate the distance travelled by the vehicle from the time the parachute opens until the vehicle stops.

> *Space for working and answer*

2

(iii) Calculate the acceleration of the vehicle while it is slowing down.

> *Space for working and answer*

2

(iv) Calculate the unbalanced force on the vehicle while it is slowing down.

> *Space for working and answer*

2

[Turn over

Marks K&U P!

9. (*c*) **(continued)**

(v) Calculate the kinetic energy of the vehicle at the instant the parachute opens.

Space for working and answer

2

10. A metal guitar string, fixed to a wooden base, is connected to an oscilloscope. A magnet is placed so that the string is between the poles of the magnet, as shown.

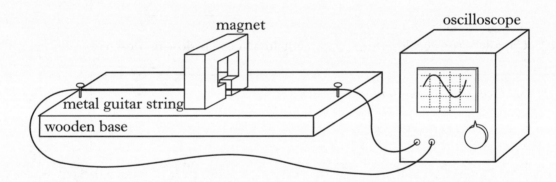

When the string is plucked, a sound is produced and a voltage is induced in the string. The induced voltage is displayed on the screen of the oscilloscope.

(*a*) (i) Why is a voltage induced when the string is plucked?

....

.... 1

(ii) State one change that can be made so that a larger voltage is induced.

.... 1

Marks | K&U | PS

10. **(continued)**

(b) The oscilloscope gain setting and trace are shown.

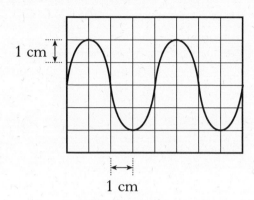

Calculate the peak voltage.

Space for working and answer

2

(c) A different metal string is used to produce a louder sound of higher frequency. No other changes are made to the equipment.

Draw a possible new trace on the blank screen below.

trace produced by original string

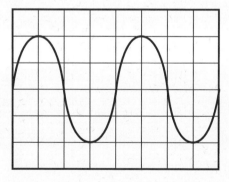

trace produced by second string

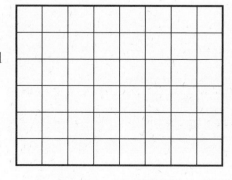

2

Marks K&U PS

11. A mass of 500 g of a substance is heated with a 30 W heater. A temperature probe is inserted into the substance.

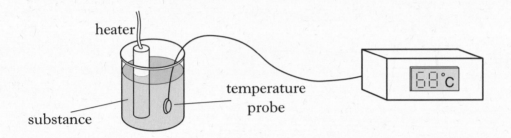

The substance is initially solid and at room temperature. The graph below shows the variation of the temperature of the substance from the time the heater is switched on.

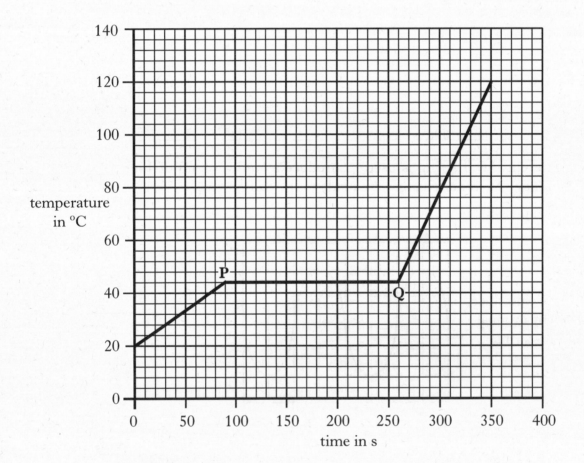

(a) State the value of room temperature.

.... 1

Marks | K&U | PS

11. (continued)

   (*b*) (i)   Why does the temperature of the substance remain constant
            between P and Q?

            ....   **1**

       (ii)   Calculate the energy transferred by the heater during the time
            interval PQ.

            Space for working and answer

                                                                                        3

       (iii) Calculate the specific latent heat of fusion of the substance.

            Space for working and answer

                                                                                        2

                                                                        [Turn over

Marks

12. A bicycle lamp contains four LEDs W, X, Y and Z and a 3 V battery. The lamp uses a pulse generator to make two of the LEDs flash. A simplified circuit diagram of the bicycle lamp is shown.

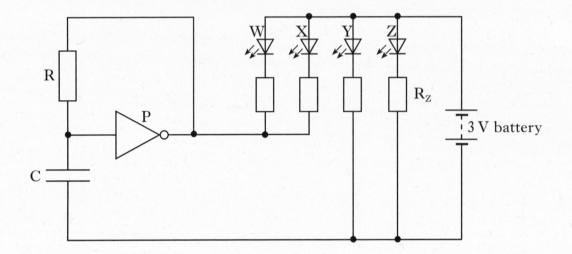

(*a*) (i) Which LEDs flash when the lamp is operating?

.... **1**

(ii) State two changes that could be made to the circuit to increase the frequency at which the LEDs flash.

....

.... **2**

(*b*) When LED Z is lit, the current in it is 15 mA and the voltage across it is 1·8 V.

Calculate the resistance of R_Z.

Space for working and answer

3

Marks | K&U | PS

13. The table below has information about three telescopes used to detect radiation from space.

objective lens	Refracting telescope in Edinburgh, with 150 mm diameter objective lens.
detector — curved reflector	Radio telescope at Jodrell Bank, with a curved reflector of diameter 76 m.
detector — curved reflector	Radio telescope at Arecibo, Puerto Rico, with a curved reflector of diameter 300 m.

(a) What type of radiation is detected by a refracting telescope?

.... **1**

(b) Why are different types of telescope used to detect radiation from space?

....

.... **1**

(c) In a radio telescope, where is the detector placed in relation to the curved reflector?

....

.... **1**

(d) Explain which of the three telescopes shown above is best for detecting very weak radio signals from deep space.

....

....

....

.... **2**

Marks

14. A space vehicle consists of a rocket engine, fuel and a probe. When sitting on the launch pad, the total mass of the space vehicle is 150 000 kg.

(a) Calculate the weight of the space vehicle on the launch pad.

Space for working and answer

2

(b) The space vehicle is launched. Shortly after lift-off, it is at a height of 650 km above the surface of the Earth. At this time, 80 000 kg of fuel have been used.

Give **two** reasons why the weight of the space vehicle is now less than it was on the launch pad.

Reason one...

....

Reason two ....

.... 2

(c) The space vehicle travels into a region of space where the gravitational field strength is zero. The engine is now switched off.

Describe and explain the motion of the vehicle.

....

....

.... 2

DO NOT WRITE IN THIS MARGIN

Marks | K&U | PS

15. Some members of the electromagnetic spectrum are named below.

TV and Radio		Infrared	Visible light		X-rays	Gamma rays

(*a*) Write the names of the missing radiations in the correct spaces in the diagram above. **2**

(*b*) State **one** radiation that has a lower frequency than visible light.

.... **1**

(*c*) State **one** detector of X-rays.

.... **1**

(*d*) State **one** medical use of infrared radiation.

.... **1**

[*END OF QUESTION PAPER*]

YOU MAY USE THE SPACE ON THIS PAGE TO REWRITE ANY ANSWER YOU HAVE DECIDED TO CHANGE IN THE MAIN PART OF THE ANSWER BOOKLET. TAKE CARE TO WRITE IN CAREFULLY THE APPROPRIATE QUESTION NUMBER.

[BLANK PAGE]

FOR OFFICIAL USE

C

K & U   PS

Total Marks

3220/402

NATIONAL
QUALIFICATIONS
2005

TUESDAY, 24 MAY
10.50 AM – 12.35 PM

PHYSICS
STANDARD GRADE
Credit Level

Fill in these boxes and read what is printed below.

Full name of centre

Town

Forename(s)

Surname

Date of birth
Day Month Year   Scottish candidate number   Number of seat

1 All questions should be answered.

2 The questions may be answered in any order but all answers must be written clearly and legibly in this book.

3 Write your answer where indicated by the question or in the space provided after the question.

4 If you change your mind about your answer you may score it out and rewrite it in the space provided at the end of the answer book.

5 Before leaving the examination room you must give this book to the invigilator. If you do not, you may lose all the marks for this paper.

6 Any necessary data will be found in the **data sheet** on page two.

7 Care should be taken to give an appropriate number of significant figures in the final answers to questions.

SCOTTISH
QUALIFICATIONS
AUTHORITY

DATA SHEET

Speed of light in materials

Material	Speed in m/s
Air	$3 \cdot 0 \times 10^8$
Carbon dioxide	$3 \cdot 0 \times 10^8$
Diamond	$1 \cdot 2 \times 10^8$
Glass	$2 \cdot 0 \times 10^8$
Glycerol	$2 \cdot 1 \times 10^8$
Water	$2 \cdot 3 \times 10^8$

Speed of sound in materials

Material	Speed in m/s
Aluminium	5200
Air	340
Bone	4100
Carbon dioxide	270
Glycerol	1900
Muscle	1600
Steel	5200
Tissue	1500
Water	1500

Gravitational field strengths

	Gravitational field strength on the surface in N/kg
Earth	10
Jupiter	26
Mars	4
Mercury	4
Moon	$1 \cdot 6$
Neptune	12
Saturn	11
Sun	270
Venus	9

Specific heat capacity of materials

Material	Specific heat capacity in J/kg °C
Alcohol	2350
Aluminium	902
Copper	386
Diamond	530
Glass	500
Glycerol	2400
Ice	2100
Lead	128
Water	4180

Specific latent heat of fusion of materials

Material	Specific latent heat of fusion in J/kg
Alcohol	$0 \cdot 99 \times 10^5$
Aluminium	$3 \cdot 95 \times 10^5$
Carbon dioxide	$1 \cdot 80 \times 10^5$
Copper	$2 \cdot 05 \times 10^5$
Glycerol	$1 \cdot 81 \times 10^5$
Lead	$0 \cdot 25 \times 10^5$
Water	$3 \cdot 34 \times 10^5$

Melting and boiling points of materials

Material	Melting point in °C	Boiling point in °C
Alcohol	−98	65
Aluminium	660	2470
Copper	1077	2567
Glycerol	18	290
Lead	328	1737
Turpentine	−10	156

Specific latent heat of vaporisation of materials

Material	Specific latent heat of vaporisation in J/kg
Alcohol	$11 \cdot 2 \times 10^5$
Carbon dioxide	$3 \cdot 77 \times 10^5$
Glycerol	$8 \cdot 30 \times 10^5$
Turpentine	$2 \cdot 90 \times 10^5$
Water	$22 \cdot 6 \times 10^5$

SI Prefixes and Multiplication Factors

Prefix	Symbol	Factor	
giga	G	1 000 000 000	$= 10^9$
mega	M	1 000 000	$= 10^6$
kilo	k	1000	$= 10^3$
milli	m	0·001	$= 10^{-3}$
micro	μ	0·000 001	$= 10^{-6}$
nano	n	0·000 000 001	$= 10^{-9}$

1. A car driver listens to a radio station broadcasting on 1500 kHz.

    (a) Calculate the wavelength of the radio broadcast.

    > *Space for working and answer*

    Marks 2

    (b) The table shows the frequency range of the different wavebands on the radio receiver.

Waveband	Frequency range
long wave	30 kHz – 300 kHz
medium wave	300 kHz – 3 MHz
short wave	3 MHz – 30 MHz
F.M.	30 MHz – 300 MHz

    From the table, write down the waveband of the radio station that the driver is listening to.

    .... **1**

    (c) A passenger in the car listens to a personal CD player.
    The car enters a tunnel.

    radio transmitter

    As the car enters the tunnel, the sound from the radio fades, but the sound from the CD player can still be heard.

    (i) Explain why the sound from the radio fades.

    ....

    .... **1**

    (ii) Explain why the sound from the CD player can still be heard.

    ....

    .... **1**

Marks | K&U | PS

2. A television receiver is used to pick up a signal from a television transmitter.

(a) The block diagram represents a television receiver.

```
aerial → [ ] → audio → audio → loudspeaker
              decoder  amplifier

              → video → video → picture tube
                decoder  amplifier
```

(i) On the diagram, label the part of the receiver that has been left blank.

1

(ii) State the purpose of the aerial.

....

....

1

(iii) One other necessary part of the television receiver is not shown on the block diagram.

Name this part.

....

1

(iv) Which part of the television receiver transforms electrical energy to light energy?

....

1

(b) In the transmitter, a video signal is combined with a carrier wave to produce a signal for transmission.

(i) Circle the correct phrase to complete this sentence.

The carrier wave has a frequency that is $\begin{Bmatrix} \text{higher than} \\ \text{the same as} \\ \text{lower than} \end{Bmatrix}$ the frequency of the video signal.

1

(ii) Why is the carrier wave needed for transmission?

....

....

1

(iii) Name the process of combining the waves for transmission.

....

1

Marks | K&U | PS

3. A student sets up the apparatus **exactly** as shown to measure the speed of sound in air.

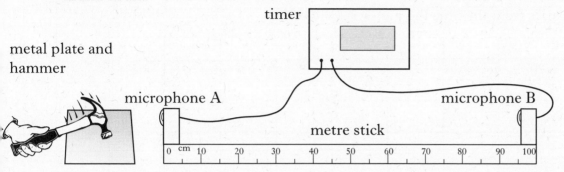

timer

metal plate and hammer

microphone A

microphone B

metre stick

0 cm 10 20 30 40 50 60 70 80 90 100

Striking the metal plate with the hammer produces a sound. Timing starts when the sound reaches microphone A, and stops when the same sound reaches microphone B.

(a) The student carries out the experiment three times and records the results shown in the table.

trial	distance between microphones (m)	time recorded on timer (s)
1	1·00	0·00287
2	1·00	0·00282
3	1·00	0·00286

Use **all** of the student's results to calculate the value of the speed of sound.

Space for working and answer

3

(b) Suggest a reason why the student's results do **not** give the value of 340 m/s for the speed of sound in air, as quoted in the data sheet.

....

.... 1

Marks | K&U | PS

4. A mains vacuum cleaner contains a motor that takes 3·0 s to reach full speed after being switched on. The graph shows how the current in the motor varies from the time the motor is switched on.

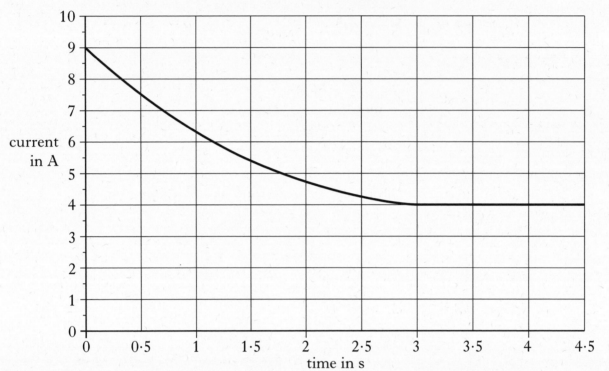

(a) (i) State the current when the motor has reached full speed.

.... **1**

(ii) Calculate the power of the motor when it has reached full speed.

Space for working and answer

3

(b) The vacuum cleaner is connected to the mains supply by a flex fitted with a fused plug.

(i) All the fuses shown are available.

| 3 ampere | | 5 ampere |
| 10 ampere | | 13 ampere |

Which one of these fuses is **most** suitable for fitting in the plug?

.... **1**

Marks | K&U | PS

4. **(*b*) (continued)**

(ii) State the purpose of the fuse fitted in the plug.

....

.... **1**

(iii) Explain why the fuse must be connected in the live wire.

....

....

.... **1**

[Turn over

5. A post office contains an emergency alarm circuit. Each of three cashiers has an alarm switch fitted as shown. Lamps come on and a bell sounds if an alarm switch is closed.

Switch P   Switch Q   Switch R

The circuit diagram for the alarm is shown.

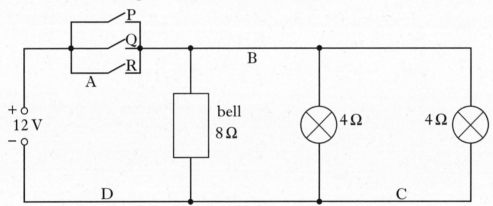

(a) The alarm circuit is to be controlled by a master switch.

Which position, A, B, C or D, is most suitable for the master switch?

1

(b) Each lamp has a resistance of $4\,\Omega$ and the bell has a resistance of $8\,\Omega$. The circuit uses a 12 V supply.

   (i) Calculate the total resistance of the alarm circuit.

Space for working and answer

2

DO NOT WRITE IN THIS MARGIN

Marks | K&U | PS

5. (*b*) (continued)

(ii) Calculate the current from the supply when the alarm is operating.

Space for working and answer

2

(*c*) Brighter lamps are fitted in the alarm circuit.

Explain how this change affects the resistance of the circuit.

....

....

.... 2

[Turn over

Marks | K&U | PS

6. In the eye, refraction of light occurs at the cornea and at the eye lens.

   (a) What is meant by refraction of light?

....

....   **1**

   (b) The diagram below shows light rays entering the eye of a short-sighted person.

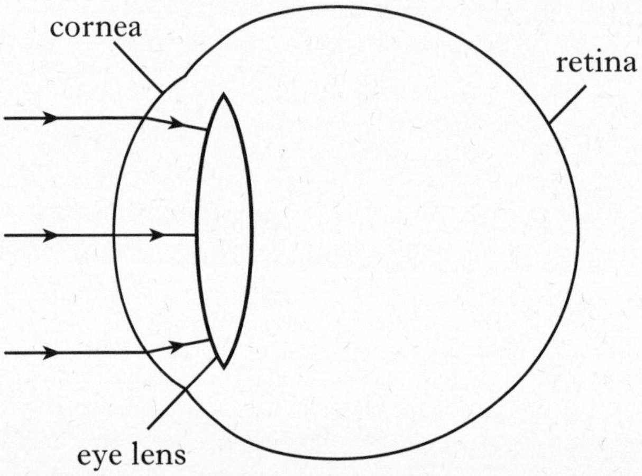

cornea

retina

eye lens

     (i) Complete the diagram above to show how the light rays reach the retina of this short-sighted eye.   **1**

     (ii) A concave lens of focal length 400 mm is needed to correct the vision in this eye.

        Calculate the power of this lens.

Space for working and answer

   2

Marks | K&U | PS

6. (continued)

(c) Short-sight can be corrected using a laser to reshape the cornea.

(i) For this treatment a pulsed laser is used. Each pulse lasts for a time of 0·2 ms and transfers 5 mJ of energy.

Calculate the power rating of the laser.

Space for working and answer

2

(ii) What effect does laser surgery have on the focal length of the cornea?

.... **1**

(iii) When a laser is in use, a warning sign similar to the one shown must be displayed.

DANGER

LASER RADIATION

Avoid eye or skin exposure

Why must a warning sign be displayed?

....

.... **1**

[Turn over

DO NOT
WRITE IN
THIS
MARGIN

Marks | K&U | PS

7. Radioactive sources are used in medical investigations.

(a) A technician uses a Geiger-Muller tube, a counter and a timer to measure the half-life of a radioactive source. The source and the tube are placed in a lead box to exclude background radiation.

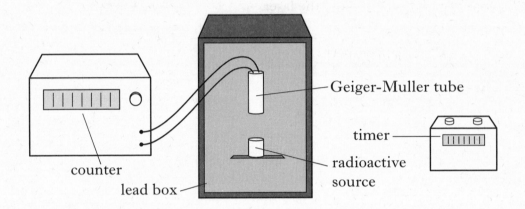

(i) Describe how the apparatus is used to measure the half-life of the radioactive source.

....

....

....

....

....

....   3

(ii) The half-life of the source is 10 minutes. The initial count rate is 1200 counts per minute.

Calculate the count rate after 40 minutes.

Space for working and answer

2

DO NOT WRITE IN THIS MARGIN

7. (continued)

Marks | K&U | PS

(*b*) Dose equivalent measures the biological effect of radiation.

(i) What unit is used to measure dose equivalent?

.... **1**

(ii) State **two** factors that dose equivalent depends on.

....

.... **2**

[Turn over

Marks | K&U | PS

8. The circuit shown is used to investigate the switching action of a transistor.

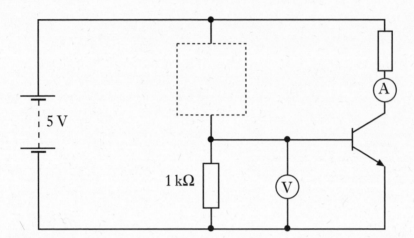

(a) Draw the symbol for a variable resistor in the dotted box in the above diagram.

1

(b) The graph shows how the ammeter reading varies with the voltmeter reading when the resistance of the variable resistor is changed.

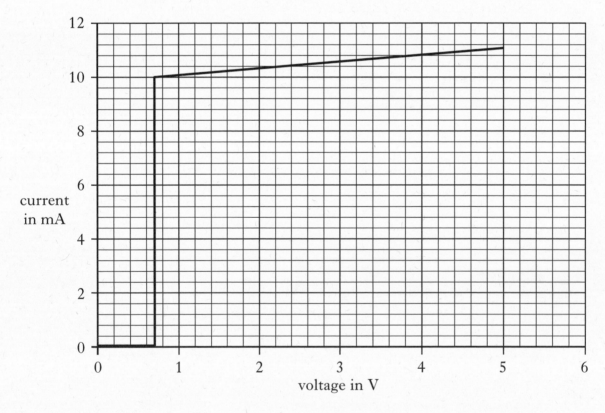

(i) State the voltage at which the transistor starts to conduct.

....

1

DO NOT WRITE IN THIS MARGIN

Marks | K&U | PS

8. (*b*) (continued)

(ii) Calculate the voltage across the variable resistor when the transistor starts to conduct.

Space for working and answer

1

(iii) Calculate the resistance of the variable resistor when the transistor starts to conduct.

Space for working and answer

2

[Turn over

Marks K&U PS

9. A machine packs eggs into boxes. The eggs travel along a conveyor belt and pass through a light gate that operates a counter. After the correct number of eggs has passed through the light gate, the counter resets and the box is exchanged for an empty one.

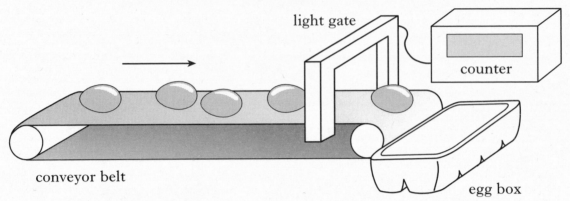

(a) The light gate consists of a light source and detector.

State a suitable component to be used as the detector.

.... 1

(b) Part of the counter circuit is shown.

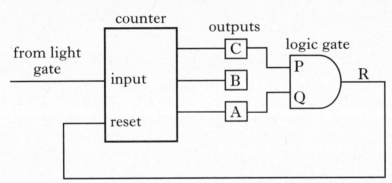

The input to the counter goes to logic 1 every time an egg passes through the light gate. When the reset to the counter goes to logic 1, the outputs go to zero.

The table below shows the logic states of the three outputs A, B and C of the counter as eggs pass the detector.

Number of eggs	A	B	C
0	0	0	0
1	0	0	1
2	0	1	0
3	0	1	1
4	1	0	0
5	1	0	1
6	1	1	0
7	1	1	1

9. **(b)** **(continued)**

(i) Complete the truth table for the logic gate shown.

P	Q	R
0	0	
0	1	
1	0	
1	1	

(ii) How many eggs are being packed into each box when the logic gate is connected to the counter outputs as shown?

....

(iii) Complete the diagram below to show how the logic gate should be connected to the counter outputs so that six eggs can be packed in a box.

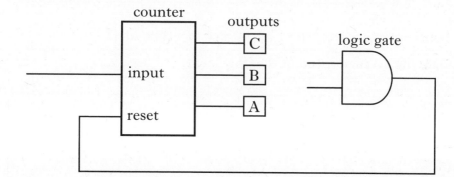

[Turn over

Marks

10. A bobsleigh team competes in a race.

(a) Starting from rest, the bobsleigh reaches a speed of 11 m/s after a time of 3·2 s.

Calculate the acceleration of the bobsleigh.

> *Space for working and answer*

2

(b) The bobsleigh completes the 1200 m race in a time of 42·0 s.

Calculate the average speed of the bobsleigh.

> *Space for working and answer*

2

(c) Describe how the instantaneous speed of the bobsleigh could be measured as it crosses the finish line.

....

....

....

....

2

DO NOT WRITE IN THIS MARGIN

Marks	K&U	PS

10. (continued)

(*d*) To travel as quickly as possible, frictional forces must be minimised.

State **two** methods of reducing friction.

....

.... **2**

[Turn over

Marks | K&U | PS

11. A train travels up a mountain carrying skiers in winter and tourists in summer.

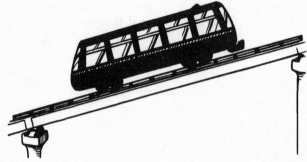

(a) The graph shows how the speed of the train varies with time for the journey in winter.

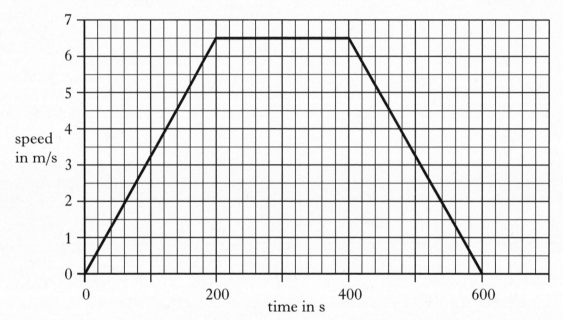

(i) Calculate the acceleration of the train during the first 200 s.

> *Space for working and answer*

2

(ii) Calculate the length of the journey.

> *Space for working and answer*

2

DO NOT WRITE IN THIS MARGIN

Marks | K&U | PS

11. (continued)

(b) The mass of the train is 15 000 kg. During the journey the train travels through a height of 460 m.

Calculate the potential energy gained by the train.

Space for working and answer

2

(c) In summer, the train takes a time of 1200 s to travel up the mountain so that tourists can enjoy the view. The acceleration and deceleration of the train remain the same as in winter. The graph below again shows the motion of the train in winter.

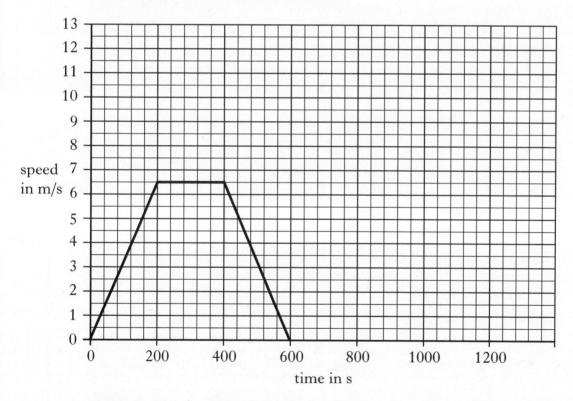

Using the axes given above, sketch a second graph showing the motion of the train in summer.
(Calculations are not required.)

2

[Turn over

DO NOT
WRITE I[
THIS
MARGIN

Marks | K&U | P[

12. An electric toothbrush contains a rechargeable battery. The battery is recharged using a transformer connected to a 230 V a.c. supply. The primary coil and the core of the transformer are sealed into the base unit. The 5 V secondary coil of the transformer is part of the toothbrush.

 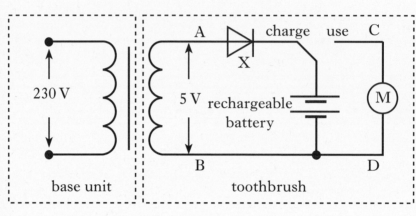

To charge the battery, the toothbrush is placed on the base unit, with the switch in the "charge" position.

(*a*) Identify the component labelled X.

.... **1**

(*b*) The primary coil of the transformer has 6440 turns.

(i) Assuming the transformer is 100% efficient, calculate the number of turns on the secondary coil.

Space for working and answer

2

(ii) When the toothbrush is charging, the current in the secondary coil is 50 mA.

(A) Calculate the output power of the transformer.

Space for working and answer

2

Marks | K&U | PS

12. (*b*)(ii) (continued)

(B) In practice, the transformer is only 40% efficient.

Calculate the current in the primary coil.

Space for working and answer

3

(iii) State **one** reason why a transformer is less than 100% efficient.

.... **1**

(*c*) Sketch the trace seen when an oscilloscope is connected across:

(i) AB when the battery is being charged;

(ii) CD when the toothbrush is removed from the base unit and the switch is in the "use" position.

Values need not be shown on either sketch.

AB   CD   **2**

[Turn over

Marks | K&U | PS

13. The apparatus shown is used to calculate the value of the specific latent heat of vaporisation of water.

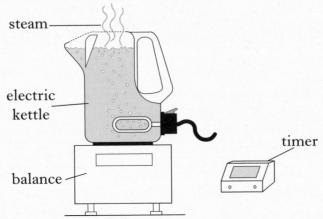

The electric kettle is rated at $3 \cdot 0$ kW. The kettle containing water is placed on the balance. The lid of the kettle is removed and the kettle is switched on. Once the water starts to boil, the kettle is left switched on for a further $85 \cdot 0$ s before being switched off.

(a) Calculate how much electrical energy is supplied to the kettle in $85 \cdot 0$ s.

Space for working and answer

2

(b) The reading on the balance decreases by $0 \cdot 12$ kg during the $85 \cdot 0$ s.

(i) Assuming all the electrical energy supplied is transferred to the water, calculate the value of the specific latent heat of vaporisation of water obtained in the experiment.

Space for working and answer

2

(ii) The accepted value for the specific latent heat of vaporisation of water is $22 \cdot 6 \times 10^5$ J/kg.

Suggest why there is a difference between this value and the value obtained in (b)(i).

....

....

1

Marks | K&U | PS

14. An astronomer uses a telescope and a camera to take a photograph of a distant galaxy.

camera   telescope   distant galaxy

(*a*) The table shows a number of lenses that are available for use in the telescope.

lens	type	focal length (mm)	diameter (mm)
P	concave	15	10
Q	convex	15	10
R	convex	1000	10
S	convex	1000	100
T	concave	1000	100

From the table, select the most suitable lenses for use as the eyepiece and the objective of the telescope.

Eyepiece ☐   Objective ☐   2

(*b*) The astronomer examines the photograph using a magnifying glass.

Complete the ray diagram to show how the magnifying glass can be used to form an image of the photograph.

Your diagram must show the position of the image.

photograph

focus   focus

3

Marks | K&U | PS

15. A spacecraft consisting of a rocket and a lunar probe is launched from the Earth to the Moon.

(a) At lift-off from the Earth, the spacecraft has a weight of 7100 kN. The thrust from the engines is 16 000 kN.

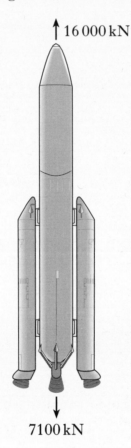

↑ 16 000 kN

↓
7100 kN

(i) Calculate the unbalanced force acting on the spacecraft.

> *Space for working and answer*

1

(ii) Calculate the mass of the spacecraft.

> *Space for working and answer*

1

Marks | K&U | PS

15. (*a*) **(continued)**

(iii) Calculate the initial acceleration of the spacecraft.

Space for working and answer

2

(*b*) As it approaches the Moon, the probe is detached from the rocket and goes into lunar orbit.

(i) While orbiting the Moon, the probe takes images of the Moon's surface. This data is sent to Earth using radio waves. The distance between the probe and Earth is 384 000 km.

Calculate the time taken for the data to reach Earth.

Space for working and answer

2

(ii) The Moon is a natural satellite and the probe is an artificial satellite.

Explain what a satellite is.

....

.... 1

(iii) The probe orbits the Moon because of the Moon's gravitational field.

Explain why the probe does not crash into the Moon.

....

....

.... 1

[END OF QUESTION PAPER]

YOU MAY USE THE SPACE ON THIS PAGE TO REWRITE ANY ANSWER YOU HAVE DECIDED TO CHANGE IN THE MAIN PART OF THE ANSWER BOOKLET. TAKE CARE TO WRITE IN CAREFULLY THE APPROPRIATE QUESTION NUMBER.

[BLANK PAGE]

Physics Credit Level 2005

1. (a) 200 m

   (b) medium wave

   (c) (i) radio waves cannot diffract into the tunnel to reach the aerial of the car radio
       (ii) CD player receives its signal/data from the CD **or** no radio waves are used **or** does not need a wave transmitter

2. (a) (i) tuner
       (ii) picks up/detects/collect all radio waves transmitted **or** within range
       (iii) power supply/ mains/ battery/ voltage supply
       (iv) (picture) tube

   (b) (i) higher than
       (ii) adds energy to the signal for transmission **or** allows single frequency transmission of audio and video **or** allows signal to be transmitted to a greater distance
       (iii) modulation/am/fm

3. (a) 351 m/s

   (b) distance used is not 1·00 m (as recorded)

4. (a) (i) 4·0 A
       (ii) 920 W

   (b) (i) 10 (ampere)
       (ii) to protect the flex
       (iii) to disconnect the appliance from the live wire **or** to disconnect the high voltage in the event of a fault

5. (a) D

   (b) (i) 1·6 Ω
       (ii) 7·5 A

   (c) (brighter lamps)→more power
       more power→more current
       more current→less resistance/lamp
       less resistance/lamp→less resistance (in circuit)

6. (a) change of speed when travelling from one medium to another

   (b) (i)

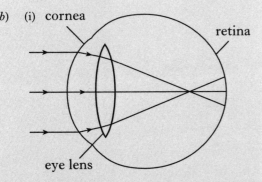

cornea
retina
eye lens

       (ii) 2·5 D

   (c) (i) 25 W
       (ii) (f) increases
       (iii) laser radiation is harmful to tissue **or** eyes

7. (a) (i) count measured for fixed time period
       several counts taken at intervals
       plot graph of count rate(-v-time)
       (ii) 75 (counts per minute)

   (b) (i) sievert **or** Sv
       (ii) Any two of:
          • type of radiation
          • type of tissue
          • weighting factor
          • quality factor
          • time (of exposure)
          • energy (absorbed)
          • absorbed dose
          • mass of tissue

8. (a)

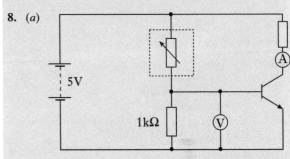

5V

1kΩ

   (b) (i) 0·7 V
       (ii) 4·3 V
       (iii) 6143 Ω **or** 6·1 kΩ

9. (a) LDR **or** solar cell **or** photodiode **or** phototransistor

   (b) (i)

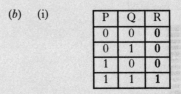

P	Q	R
0	0	**0**
0	1	**0**
1	0	**0**
1	1	**1**

       (ii) 5
       (iii)

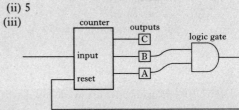

counter
outputs
logic gate
input
reset

10. (a) 3·4 m/s²

    (b) 28·6 m/s

    (c) measure length of bobsleigh
        set up a light gate and timer

        $$speed = \frac{distance}{time}$$

    (d) Any two of:
        • streamlining/aerodynamic shape
        • lubrication
        • less weight/less mass
        • smooth surfaces
        • using rollers

8. (a) NOT gate (or inverter)

(b)

Liquid Level	P	Q	R
Overfilled	0	1	0
Acceptable	1	0	0
Underfilled	1	0	1

(c) OR gate

9. (a) Average speed – the time is measured over a long distance.

(b) 9·09 s

(c) (i) Air resistance/friction/drag is increased
   (ii) 16 500 m
   (iii) −1·47 m/s^2
   (iv) 4400 N
   (v) 7·26 × 10^7 J

10. (a) (i) The wire is moving in a magnetic field.
   (ii) Any one from:
   • Stronger magnet
   • Pluck harder
   • Pluck more at right angles to the field
   • Make the wire tighter

(b) 0·4 mV

(c)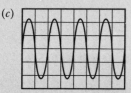

11. (a) 20 °C

(b) (i) Substance is changing state
   (ii) 5100 J
   (iii) 10 200 J/kg

12. (a) (i) W and X
   (ii) Any two from:
   reduce C, reduce R, increase V.

(b) 80 Ω

13. (a) Light

(b) Different telescopes are needed to detect different types of radiation

(c) At the focal point

(d) Arecibo, because it is a radio telescope and has the largest diameter reflector.

14. (a) 1 500 000 N

(b) 1. Mass of the vehicle is reduced.
   2. Gravitational field strength is less.

(c) Constant speed in a straight line, because no (unbalanced) forces act on it.

15. (a)

TV & Radio	microwave	Infrared	Visible light	ultraviolet	X-rays	Gamma rays

(b) Any one from: TV/radio, microwave, infrared

(c) Photographic film

(d) Any one from: Thermographic images, repair muscle tissue, detecting tumours, measuring blood pressure, measuring temperature, sealing blood vessels

Physics Credit Level 2003 (cont.)

13. (*a*) $P = \dfrac{E}{t} = \dfrac{mgh}{t}$ $t = 60$ s

$P = \dfrac{6000 \times 10 \times 5}{60}$

= 5000 (W)

(*b*) (i) efficiency = 0·598
(ii) Any two from:
  • friction in bearings/wheel/generator
  • heating in generator
  • resistance in wires
  • splashing/water loss (from buckets)
(iii) $t = 700$ s
(iv) Any one from:
  • not all the heat is transferred to the air
  • movement of air (so greater volume heated)
  • heating (the fabric of) the barn
  • heat loss to the environment

14. (*a*) electromagnetic spectrum

(*b*) (i) Gamma: sterilising (medical instruments)
  Ultraviolet: tanning (with a sun-ray lamp)
  Infrared - treating injuries (using a heat-lamp) OR linking (networked) computers (through optical fibres)
(ii) longest wavelength: infrared
  highest frequency: gamma

15. (*a*) $t = 0·18$ s

(*b*) constant velocity (speed) horizontally and vertical/downwards acceleration (caused by gravity)
OR
no unbalanced force horizontally and vertical/downwards force (of gravity)

(*c*) $d = 0·162$ m (= 16·2 cm)

Physics Credit Level 2004

1. (*a*) (i) 900 MHz
  (ii) 0·2 m

(*b*) (i) Any two from:
  • Less amplification needed
  • More signals (per fibre)
  • Less energy loss
  • Fewer repeater stations
  • No (electrical) interference
  • Better quality
  • Cost
  • Security
  • Smaller
(ii) (A) $2·0 \times 10^8$ m/s
  (B) $2·4 \times 10^5$ m

2. (*a*) 1500 images

(*b*) **separate** the signal from the carrier.

(*c*) (i) The three primary colours are needed to produce all the colours seen.
(ii) Mask stops beams overlapping so that each beam hits the correct colour.

3. (*a*) To store/supply energy

(*b*) (i) Total distance moved by force
  $= 500 \times 400 \times 10^{-3} = 200$ m
  WD $= F \times d = 9·0 \times 200 = 1800$ J
(ii) (A) 1620 J
  (B) 2160 s

4. (*a*) Sidelights

(*b*) With S2 open: relay coil is de-energised, causing the relay switch to open, the headlight circuit is broken. The sidelights are not controlled by the relay so they stay on.

(*c*) (i) 2 A
(ii) 1200 C
(iii) Headlight is a higher power rating than the sidelight, so more energy is transformed each second.

5. (*a*) (i) 400
  (ii) 8 Ω

(*b*) When the push-switch is pressed, the electromagnet is energised. The iron bar is attracted inside the electromagnet freeing the door.

6. (*a*) Can **only** see distant objects clearly OR Cannot see close objects clearly

(*b*) Left Right

(*c*) 0·4 m

7. (*a*) (i) Gain/loss of charge/electrons from an atom.
(ii) electrons/charged particles move (to form a current)

(*b*) Alpha radiation causes more ionisation

(*c*) becquerel or Bq

5. (*a*)

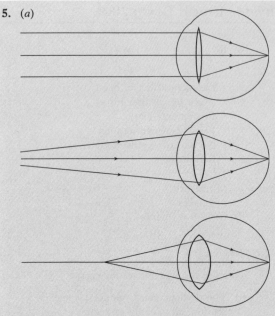

(*b*) 40 D

6. (*a*) 3 A

(*b*) $V = 6$ V

(*c*) $R_y = 6\ \Omega$

(*d*) more current/double current/current = 6 A so total resistance must be reduced/halved

7. (*a*) Time taken for the activity/no. of radioactive nuclei to reduce by one half (of the original value/number)

(*b*) (i) Alpha radiation would (all) be absorbed by the paper
OR
Alpha radiation has too short a range

(ii) R
Beta radiation not completely absorbed (by paper)
Long half-life (for continuous operation)

(*c*) To absorb radiation given out in other directions (than towards paper)/safety/to protect workers

(*d*) 4 hours

8. (*a*) AND

(*b*)

	Output R
1	
0	————
1	
0	————
1	
0	————
1	⊓⊓
0	

(*c*) (i) (C discharged, so) X is logic 0, Y is logic 1
C charges (through R)/voltage at X increases (to logic 1) so Y goes to logic 0/
C discharges (through R) so X goes to logic 0, Y to logic 1/repeats

(ii) Increase R OR increase C OR increase R and C

9. (*a*) variable resistor

(*b*) (i) transistor
(ii) (electronic/voltage controlled) switch

(*c*) As the temperature changes, the voltage across R/ at the base (of the transistor)/at the junction of P and R changes.
Increase of voltage (at the base of the transistor) causes the transistor to switch on
OR
Decrease of voltage (at the base of the transistor) causes the transistor to switch off.

10. (*a*) (i) first
(ii) second

(*b*) 14 m

(*c*) $a = -4 \cdot 5$ m/s^2

11. (*a*) (i) 3 N
(ii) $a = 0 \cdot 5$ m/s^2

(*b*) constant/uniform/steady speed because the forces are balanced

12. (*a*) step-down

(*b*) $n_s = 240$ (turns)

(*c*) (i) (A) $P_{out} = 64 \cdot 86$ W
(B) $I_p = 0 \cdot 3$ A
(ii) Any one from:
• power loss due to heating in coils
• power loss due to resistance of wires/coils
• heating in core due to eddy currents
• power/energy loss due to heat/vibration/ sound generated (in the transformer)
• power loss due to hysterisis/magnetising core

Physics Credit Level 2002 (cont.)

9.   (a)   4200 J
    (b)   (i) Q
          (ii) 1·8 m
          (iii) energy is transferred (as heat)
              due to (the force of) friction
              or energy is lost to the system
              or work done against friction

10.   (a)   0·5 s
    (b)   2·0 m/s^2
    (c)   (i) 240 m   (ii) 28·75 m (29 m)

11.   (a)   300
    (b)   (i) 4·5 A   (ii) 0·23 A
    (c)   (i) $P = I^2 R$ **or** $V = IR$

$$\therefore R = \frac{P}{I^2} \qquad\qquad \therefore R = V/I$$

$$= \frac{18}{1 \cdot 5}2 \qquad\qquad = \frac{12}{1 \cdot 5}$$

$$= 8 \cdot 0 \,\Omega \qquad\qquad = 8 \cdot 0 \,\Omega$$

          (ii) 2·7 Ω

12.   (a)   15 120 (J)
    (b)   (i) 995 J kg^{-1} °C^{-1}
          (ii) (A) not all of the energy is transferred as heat to the block
              (B) lag/insulate the aluminium block

13.   (a)   weight per unit mass
          or pull of Earth ⎫
              force of gravity   ⎬ per ⎰ unit mass
              force due to   ⎭   ⎱ kilogram
              gravitational field

    (b)

Stage	Gravitational field strength (N/kg)	Mass (kg)	Weight (N)
on the Moon	1·6	21	**33·6**
at a point during the journey	0	**21**	**0**
on the Earth	10	**21**	**210**

Physics Credit Level 2003

1.   (a)

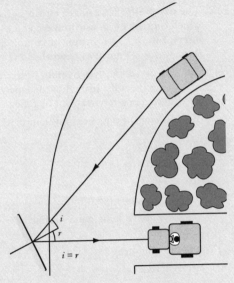

    (b) reversibility of rays
        OR mirrors can work in both directions/ways
        OR rays can go backwards/other way

2.   (a) $v = 1 \cdot 2$ m/s
    (b) $f = 0 \cdot 25$ Hz
    (c) $\lambda = 4.8$ m
    (d) $v = \dfrac{d}{t}$

        wave travels $d = 1\lambda$ in 1 period (T)

        so $v = \dfrac{\lambda}{T}$ but $f = \dfrac{1}{T}$ so $v = \lambda f$

        OR
        frequency is number of waves/second
        wavelength is length of 1 wave
        $\Rightarrow f \times \lambda =$ "length" of waves per second $= \dfrac{d}{t}$

3.   (a)   (i)   230 V
          (ii)   parallel

    (b) $I = 0 \cdot 74$ A

    (c) $R = 481$ Ω

    (d) To protect the flex/(multi-way) adaptor OR fuse melts instead of flex OR stop flex overheating

4.   (a) (1 or 5) and (2 or 4) and 3 OR red + green + blue
    (b) (1 or 5) and 3 OR green + red
    (c) 2 or 4 OR blue

Physics Credit Level 2002

1. (a) The transmitter transmits a **radio** signal, which consists of an **audio** wave and a **carrier** wave. The process of combining these waves is known as **modulation**.

(b) (i) Any correct answer relating to signal strength—hills/diffraction/distance/interference/far away/out of range

(ii) (different) frequency/wavelength

2. (a) (i) 3×10^8 m/s

(ii) $2 \cdot 8 \times 10^{-3}$ s

(iii) $2 \cdot 2$ m

(b) period 24 hours/1440 minutes

so always above same point on Earth/geostationary

(c) 100/101 (minutes)

(d) infrared/IR

(e) (the) Moon

3. (a) (i) $8 \cdot 3 \, \Omega$

(ii) resistance is constant since the graph is a straight line through the origin

or since V and I vary universally

(b) (i) not a straight line graph/not constant gradient $\frac{V}{I}$ is not constant/R increases as I

(ii) (A) $3 \cdot 2$ A

(B) $38 \cdot 4$ W

4. (a) (i) (circuit) Y

(ii) Any **two** from

thinner wire/less current per cable/convenience (of adding new sockets)/less heat/cost/safety/less voltage drop

(b) lighting circuit is simple parallel—because lower current

or lighting circuit supplies (fixed) lights not sockets—separate circuits

or lighting circuit uses thinner cables—lower current

or ring circuit has two paths—and explanation similar to (a)(ii)

or different fuse value—because of different currents

(c) (i) larger current/lot of energy/more power

(ii) 15 840 000 (J)

(d) (i) safety or an implication of safety eg prevent electrocution

(ii) casing live (because of fault); earth wire gives low resistance path/large current; fuse blows; appliance isolated from supply

5. (a) (i) (sounds of) $f > 20\,000$ Hz

or sounds above upper frequency/pitch value

(ii) $1 \cdot 25 \times 10^{-5}$ s

(b) (i) (ultra) sound reflects off baby (in womb) reflected (ultra) sound is picked up (by receiver)

(ii) ultrasound does not damage cells

or X-rays can damage (living) cells

or ultrasound is not **ionising** radiation

or X-rays are **ionising** radiation

6. (a) (i)

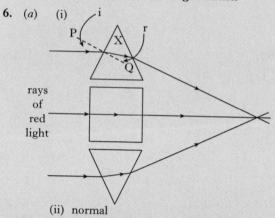

(ii) normal

(iii) may also be shown on bottom prism—**must** include normal

(b) convex (converging)

7. (a) 1 V

(b) (i) AND

(ii) OR

(iii)

P	Q	R	S	T
0	0	0	0	0
0	1	0	0	0
1	0	0	0	0
1	1	1	0	1
0	0	0	1	1
0	1	0	1	1
1	0	0	1	1
1	1	1	1	1

(iv) to raise the barrier in an emergency/if LDR or pay machine circuit faulty/no money/no change

8. (a) loudspeaker

(b) filament lamp

Any **one** from

greater light output/white light/LED is a low current device

(c) (i)

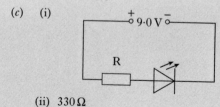

(ii) $330 \, \Omega$

Physics Credit Level 2001 (cont.)

8.   (a)   (i)   Power gain = 4000

             (ii) $V = 24\,V$

     (b) $R_T = 4 \cdot 5\,\Omega$

     (c)   256 Hz

9.   (a)   (i) $\bar{v} = 15 \cdot 2\,m/s$

             (ii)   Instantaneous speed is over a (very) small time interval/is (always) changing

                Average speed is taken over a long time interval

     (b)   (i) $E_k = 8410\,J$

             (ii) $F = 168 \cdot 2\,N$

10.   (a)   (i) $a = 1 \cdot 5\,m/s^2$

             (ii) $F = 402\,000\,N$

             (iii)   greater during 10 – 40 s because the gradient of the v–t graph is greater **or** acceleration is greater

             (iv)   distance (length of runway)

                  = area under graph

$$= (\tfrac{1}{2} \times 10 \times 15) + (30 \times 15)$$
$$\quad + (\tfrac{1}{2} \times 30 \times 65)$$
$$= 75 + 450 + 975$$
$$= 1500\,m$$

     (b)   (i)   The engine thrust is **greater than** the air friction force.

             (ii)   The lift is **equal to** the weight.

11.   (a)   (i)   Fossil fuel: Waste is not radioactive

                Nuclear: More energy/ kilogram of fuel **or** Small mass of waste produced

             (ii)   Coast or river

                Both need (a large mass of) cooling water

     (b)   (i)   Nuclear → heat

             (ii)   Kinetic → electrical

     (c)   (i)   Stage 1: A uranium nucleus is bombarded by a **neutron**.

                Stage 2: The uranium nucleus disintegrates, producing fission fragments, two **neutrons** and **heat**.

             (ii)   Each fission reaction produces (two) neutrons

                These can cause further fissions.

                More neutrons are produced from these fissions.

12.   (a) $E_h = 4008\,J$

     (b)   From the water/drink

     (c)   (i) $E_h = 2508\,J$

             (ii)   Any temperature in range $12 \cdot 0\,°C \rightarrow 15 \cdot 0\,°C$

                Because less heat is transferred to the contents from the surroundings

                or heat to melt ice comes from water only

13.   (a)   X-rays are absorbed by the (Earth's) atmosphere

     (b)

Gamma rays	X-rays	Ultraviolet	(Visible) light	Infrared	Microwaves	Radio waves

     (c)   Different signals have different wavelengths.

           Different wavelengths need different types of detectors.

     (d)   Q is fired (for a short time) then switched off.

           P is fired (for the same time) then switched off.

Pocket answer section for
SQA Credit Physics
2001 – 2005

© 2005 Scottish Qualifications Authority, All Rights Reserved
Published by Leckie & Leckie Ltd, 8 Whitehill Terrace, St Andrews, Scotland, KY16 8RN
tel: 01334 475656, fax: 01334 477392, enquiries@leckieandleckie.co.uk, www.leckieandleckie.co.uk

Physics Credit Level 2001

1. (a) (i) 1500 m/s
       (ii) depth = 150 m
       (iii) $\lambda = 0.05$ m

   (b)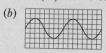
       Transmitted

       amplitude less
       wavelength same
       Reflected

   (c) Time interval unchanged because the distance **or** the speed is unchanged

2. (a) $I = 0.025$ A (b) 100 mA

3. (a)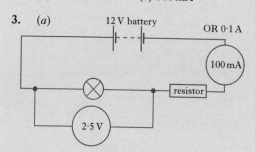

   (b) (i) $V_R = 9.5$ V
       (ii) $R = 95\ \Omega$

4. (a) Swap/reverse/change round/ change over the battery (connections)/ current
       Swap/reverse/change round/ change over the magnet (or magnetic field)/poles of magnet

   (b) (i)

       (ii) (A) smoother rotation/ operation/more even motion/more powerful/increases rotating force/ease of starting

           (B) easier to control/can shape field easier/more compact/ less mass/field stronger/can be switched off/ use on a.c. or d.c./permanent magnets can lose strength/can be reversed

5. (a) (i) Any **one** from:
           can only see far away objects clearly **or** can not focus on near objects **or** image formed behind retina **or** can not see near objects clearly

       (ii) $f = 0.4$ m

   (b)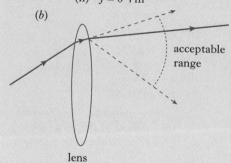
       acceptable range
       lens

6. (a) It takes 5730 years for the activity to reduce to half its original value **or** a "stock" half-life definition

   (b) (i) becquerel **or** Bq
       (ii) Radiation causes flashes (of light) These flashes are counted
       (iii) ionisation **or** fogs photographic film
            or kills (living) cells
            or sterilisation
            or changes (nature of) living cells

7. (a) (i) $R_{Th} = 700\ \Omega$
       (ii) (A) 80 °C
           (B) (If $R > 4300\ \Omega$), then $R_{Th} > 700\ \Omega$ (to maintain switching $V = 0.7$ V) so switching temperature decreases

   (b) (i) (npn) transistor
       (ii) As temperature falls:

       input $\begin{cases} R_{Th} \text{ increases} \\ V_{Th} \text{ increases} \end{cases}$

       process $\begin{cases} \text{(at } 0.7 \text{ V) transistor} \\ \text{switches on current} \\ \text{in relay coil} \end{cases}$

       output $\begin{cases} \text{closes relay switch} \\ \text{completes heater circuit} \end{cases}$

11. (*a*)　(i) 0·0325 m/s²
　　　　　(ii) 2600 m

　　(*b*) 6·9 × 10⁷ J

　　(*c*)

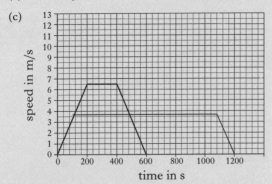

12. (*a*) diode

　　(*b*)　(i) 140
　　　　　(ii) A　　0·25 W
　　　　　　　B　　0·0027 A **or** 2·7 mA
　　　　　(iii) loss + location e.g. heat in windings,
　　　　　　　　currents in core

　　(*c*)　(i)

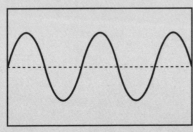

AB

　　　　(ii)

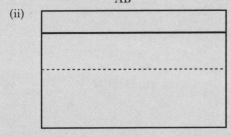

CD

13. (*a*) 2·55 × 10⁵ J

　　(*b*)　(i) 2·13 × 10⁶ J/kg
　　　　　(ii) water splashing out

14. (*a*) Eyepiece　Q
　　　　　Objective　S

　　(*b*)

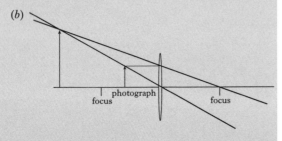

15. (*a*)　(i) 8900 kN
　　　　　(ii) 710 000 kg **or** 7·1 × 10⁵ kg
　　　　　(iii) 12·5 m/s²

　　(*b*)　(i) 1·28 s
　　　　　(ii) an object that orbits another object
　　　　　(iii) forward speed means it continually misses
　　　　　　　　the Moon while falling **or** it is in projectile
　　　　　　　　motion